POSSUM,
CLOVER
& HADES

Schiffer Military/Aviation History
Atglen, PA

The 475TH *Fighter Group in World War II*

John Stanaway

Dust jacket, and color profile artwork by Steve Ferguson, Colorado Springs, CO

DEAN'S DUTY
See page 58 for details on the artwork and mission.

Acknowledgements

The core of this history is established on the work of Norbert Krane and Carroll R. Anderson. Before he died Krane amassed an impressive collection of photos and other documents relating to the 475th Fighter Group. His estate turned the file over to me in the hope of producing published work on the Fifth Air Force. I had already decided to carry on the effort of my generous correspondent, Bob Anderson, in his effort to produce a definitive history of the 475th. Bob had died in the midst of his growing writing career, and I felt the necessity of continuing his desire to see the history written.

Another person who is responsible for the 475th history coming into realization is Dennis Glen Cooper. The venerable Intelligence Officer of the 431st Fighter Squadron and later the entire group kept me supplied with notes about the pilots and crews that he knew as well as the quality photos for which he is so famous. Many of the photos in this book that come via other sources originated with Cooper.

Other old friends jumped in enthusiastically as soon as they heard of the project. Those who provided information or photos include Jack Cook, Dwayne Tabatt, Jim Crow, Jeff Ethell, Bill Hess, John Campbell, Mark Copeland, Steve Blake, Jim Lansdale, Bob Rocker, Henry Sakaida, Bruce Hoy, and Osamu Tagaya.

Other 475th Fighter Group veterans who supplied information include dozens of anonymous people who sat in on countless free conversations or submitted to interviews of one kind or another. Those veterans who can be identified include:

Headquarters, 475th FG
Colonel Charles H. MacDonald
Colonel John S. Loisel
Dennis Glen Cooper

431st FS
Harry W. Brown
Fredric Champlin
Darren Champlin
Marion F. Kirby
Louis D. Du Montier
Vincent T. Elliott
John A. Tilley
Harold W. Gray
Harold N. Madison
Melvin J. Allan
Tony Paplia
George E. Jeschke
Seymour V. Prell
Robert Werth

432nd FS
Grover D. Gholson
James C. Ince
Perry J. Dahl
Joseph M. Forster
Robert Maxwell
Robert Schuh
Vincent Steffanic
Rozell A. Stidd
Dean T. Dutrack

433rd FS
Joseph T. McKeon
John K. Parker
Warren R. Lewis
Martin L. Low
John S. Babel
William G. Jeakle
Robert M. Tomberg
John Purdy
Clarence J. Rieman
Robert W. Schick
Calvin C. Wire
William L. Hasty
Ted W. Hanks
Richard E. Broski
George W. Rath

Book Design by Robert Biondi

First Edition
Copyright © 1993 by John Stanaway.
Library of Congress Catalog Number: 93-84493

Printed in the United States of America.
ISBN: 0-88740-518-5

We are interested in hearing from authors with book ideas on related topics.

Published by Schiffer Publishing Ltd.
77 Lower Valley Road
Atglen, PA 19310
Please write for a free catalog.
This book may be purchased from the publisher.
Please include $2.95 postage.
Try your bookstore first.

Contents

Foreword

Sometimes things seem to fit as though made to order. Such was the case with the P-38 Lockheed "Lightning" fighter aircraft, the 475th Fighter Group, and the Southwest Pacific area. We in the 475th felt very lucky indeed to have all three squadrons in the group equipped with the P-38. The aircraft had speed, range, and firepower, and lucky for some of us had two engines. When most of your missions require long flights over open ocean, jungle, swamps and mountains you appreciate the two fans.

My good friend John Stanaway ably tells the trials, tribulations and triumphs of the 475th Fighter Group. What I can tell you is that we were totally dedicated to the mission assigned to us, and were completely convinced that the course our country was taking was not only proper, but necessary.

In case there is another real war, I hope the country will respond the same.

Charles MacDonald

Colonel Charles H. MacDonald, (*USAF Ret.*)

Dedication
To the memory of Carroll R. Anderson and Norbert Kane –
and to all those who honor the spirit of the 475th Fighter Group

Introduction

Writing the history of the 475th Fighter Group is like telling a vintage anecdote; there is more surprise in the way it is told than in its substance. Most World War II aviation buffs are aware of the manner in which the 475th was formed from the heart of V Fighter Command squadrons, how it entered combat with spectacular victories in mid-1943 and how it produced some of the highest scoring American aces of the war. But few are aware of the details of the group's struggle and success in forging *esprit de corps* or of the herculean effort and saving humor of the dedicated ground crews.

It was my privilege to talk to many 475th pilots and ground personnel over the course of about twelve years of research. The one thing that all of them share is a pride in their unique, crack fighter group and an awareness of the contribution of each man to its success. Colonel Charles MacDonald has added that the element of luck is also present; it was the right time and place, with the right people and, finally, the right fighter type in the P-38.

The P-38 was certainly effective in the fighter squadrons of the Fifth Air force, but it found a special deadliness in the hands of 475th crews. Colonel John Loisel was keen on getting proficiency in his 432nd Squadron pilots in 90 degree deflection shots without "spraying lead all over the air." He honed his own shooting skill and passed it on to his pilots, and they made the most of the terrific firepower of the P-38 in combat with the redoubtable Zero and incredibly agile Nakajima Ki-43 "Oscar."

Among all the American fighter types of World War II the P-38 depended on the training and mental attitude of its pilots for consistent success. Those groups with inspired leadership and dedicated crews excelled in combat while other units fell into mediocrity. The 475th was outstanding as a fighter unit and its personnel have universally praised the superlative merits of the P-38 Lightning.

So, then, it is a combination of men and machine that made the 475th great. I have tried to portray their successes in light of their sacrifices to reach a common goal. From the irascible Tom McGuire who became the top ace of the group, to the lusty but good humored men of the line, they all made the commitment and paid their inevitable dues to defeat an enemy who seemed so formidable at the time.

"The whole 475th Fighter Group at Amberly Field, Australia, July, 1943." (Dennis Glen Cooper)

Chapter I

MAY 1943

Sergeant Mel Allan of the 8th Fighter Squadron had just received orders transferring him into another unit. He had been quite content with his life as an aircraft mechanic at Dobodura near the northeast coast of New Guinea and resented being uprooted for some unspecified reason. What had he done? Whom had he offended to be 'Shanghaied' out of his outfit where he was doing work he loved to the satisfaction of his superiors?

The more he thought about it the more dejected he became. He was alerted with several other ground crewmen to move on a moment's notice and nobody seemed to know what the fuss was all about. For all he knew he could be spending the rest of the war on guard duty or driving a truck. He kicked down the dusty road with head low and hands tucked unmilitarily in his pockets at the prospect of his immediate future.

At the same time over in the 40th Squadron down in Port Moresby several pilots were being told that they would be converting from the P-39 that they currently flew to the P-38s of a brand new squadron. First Lieutenant John Babel was overjoyed at the news, not because he wanted to leave his old unit, but that flying the P-39 in combat was a frustrating experience and the P-38 was already enjoying a legendary reputation.

Major Martin Low, another 40th Squadron pilot and a comrade of Babel from their days with the 15th Fighter Group in Hawaii, was chosen as the first commander of the new 433rd Fighter Squadron. He would go on to further leadership positions in England with the 20th Fighter Group later in the war where he commanded the 55th Fighter Squadron and won the Distinguished Flying Cross among other decorations.

Even though he was happy to take command of the new unit Low apparently didn't have the mature skills that he later displayed in the 55th Fighter Squadron in regard to relationships with his personnel. That is, judging from the comments of 433rd veterans he picked the wrong people for the wrong jobs in squadron administration and sent the most unfortunate messages to the enlisted men by withholding deserved praise and promotions as a type of negative incentive.

The record of the 433rd Squadron was always exemplary, but the morale of the unit suffered in comparison with the other two units of the parent group by a palpable measure. If Major Low deserves praise for making the 433rd operational, he also must assume at least a portion of the criticism for initial morale problems. To make appearances even worse, his successor as commander of the 433rd was Captain Danny Roberts who immediately gained a phenomenal reputation as a leader.

If Low had suffered in popularity with the enlisted men of the 433rd, Major George Prentice garnered some negative feelings with the pilots of the 39th Fighter Squadron that he commanded.

General George Kenney (seated in center of photo) was the great proponent of the P-38 fighter and wanted a full group equipped with the type to facilitate his plans for the offensive against the Japanese in New Guinea.

Prentice was fully willing to enter into aerial combat, but was outclassed as a combat leader by some of the pilots under him. At least once his pilots were obliged to save him from disaster in what had started as an advantageous engagement for the squadron.

He was selected to lead the newly formed P-38 group and, conversely to Martin Low's experience, improved with higher command. Prentice was an organizer and administrator rather than a combat leader, using his best skills to shape and inspire the formative period of the fighter group he commanded.

Prentice took five or six freshly trained P-38 pilots who had recently been assigned to the 39th with him to help fill the roster of the 433rd Squadron. Lieutenants Charles Grice, William Jeakle, John Parkansky (later changed to Parker) and John Ehlinger were some of the pilots who went to the 433rd, and would later be involved in the squadron's combat operations.

Lieutenant Arthur Peregoy was another 39th P-38 pilot who went over to the 432nd Fighter Squadron. Most of the original cadre for that unit came from the 8th Fighter Group. The 80th Fighter Squadron had already seen combat with the P-38 and contributed Frank Tomkins as 432nd commander, along with such luminaries as James Ince, Zach Dean and Noel Lundy. Danny Roberts also came from the 80th and went to the 432nd before he took command of the 433rd.

Other pilots from the 8th Fighter Group who went into the 432nd included Vivian Cloud, Grover Gholson and John Loisel, the latter officer rising to command the 475th during the last days of the war. All three pilots would become fighter aces during their time with the 432nd. Gholson, in fact, had already claimed one Japanese aircraft shot down while he was flying a P-39 with the 36th Fighter Squadron in May, 1942.

Amberly Field – the first home of the 475th FG – in June 1943. (Harry Brown)

Most of the personnel in the 431st Fighter Squadron came from the 49th Fighter Group, as did Sergeant Allan. The 49th Group had begun operations from Australia early in 1942 and had incorporated the second P-38 squadron to see action in New Guinea when the 9th Fighter Squadron converted to Lightnings in the first days of 1943.

Major Frank Nichols already had four Japanese aircraft to his credit while flying P-40s before he left the 49th Group to take command of the 431st. Jack Mankin, Harry Brown, Art Wenige, David Allen and John Hood also came to the 431st from the 49th Group. All of them had already scored aerial victories and most would become aces with the high-scoring 431st.

Captain Verl Jett also transferred from the 8th Fighter Group to the 431st. He would take over command of the squadron after Nichols went on rotation home in November 1943. Jett had served with the 15th Fighter Group in Hawaii during the time of the Pearl Harbor attack and was one of several pilots from that unit to join the new P-38 group. Transferring almost directly from the 15th Group to the 431st were Lieutenants John Cohn, John Knox and Paul Morris.

Captain Bill Waldman and Lieutenants Billy Gresham and Howard Hedrick transferred from the 15th Group to the 432nd Squadron, while Lieutenants Ralph Cleage and Bob Tomberg went to the 433rd Squadron from the 15th Group. Lieutenants Ed Czarnecki and Marion Kirby from the 80th Fighter Squadron deviated the route a bit and ended up in the 431st where they both would become fighter aces.

Group Headquarters was augmented by Major Al Schinz of the 41st Fighter Squadron who assumed duties as executive officer and Captain William Haning of the 49th Fighter Group who took over as Operations Officer of the new group. In all, about half of the group personnel came directly from Fifth Air Force fighter squadrons. The new fighter group would be activated on May 14, 1943 and constituted the next day, but the buzz had started much earlier about the nature of all this shuffling of personnel.

The P-38 Lockheed Lightning fighter had been introduced into the Fifth Air Force during the last months of 1942. It was not a proven combat airplane yet, and fostered some degree of doubt because of its twin-engine configuration. Allied pilots were well aware of the maneuverability of the Zero and tended to think that their own fighters should either be a bit faster or much more agile.

Those pilots in Curtiss P-40 fighters became fiercely loyal to their mount since it was the most effective of the earliest types of American aircraft to be deployed. Although it was not as fast nor did it turn as well as the Zero, it did have a good diving speed and rugged construction, coupled with devastat-ing armament (six .50 caliber guns in P-40E and subsequent models).

Throughout the war the Japanese considered the P-40 to be the least effective of American fighters since their Zeros could out fly and outrun it. The P-39 was the other American fighter in the New Guinea area at the time and did not draw much greater praise from its Japanese adversaries, but at least it could keep up with the Zero at sea level. The only chance for the P-40 seemed to be having enough altitude to either attack or escape in a dive.

However it may have been seen through the eyes of its enemies, the P-40's pilots flew with a sense of confidence and scored some impressive victories. The men in the P-40 cockpit were sure that a little weight reduction and better high-altitude engine with a supercharger would make the Curtiss fighter the best in the world. Their disappointment was high when the even heavier P-38 arrived instead.

General George Kenney had assumed command of the newly formed Fifth Air Force by September 1942 and was the greatest champion of the P-38 that the U.S. Army Air Forces had. He quickly became known as an operator who was able to pick the best people for any job at hand and to inspire his subordinates to accomplish his will.

The P-40 people complained that the P-38 was simply too heavy to remain in the combat area with the more flexible Zero, that it landed with less safety than the P-40 and required more maintenance and fuel. Kenney was fully aware of the technical drawbacks plus a few that some other people did not appreciate, namely that the P-38s he was receiving had troubling problems like cooling system leaks, and that he was going to have the devil's own time finding pilots and mechanics who were well-trained enough to work with the fighter!

Kenney had sold himself on the P-38 when he commanded the Fourth Air Force in California during the spring and summer of 1942. The performance advantages over the P-39 and P-40 were convincing enough. Simply stated, those

Lt. Howard Hedrick by P-38H-1 42-66573 at Amberly Field. (Krane Files)

John Cohn at an awards ceremony later in the war. (Krane Files)

Warren Cortner at the same awards ceremony. (Krane Files)

performance advantages meant that the P-38 could fly higher, faster and, especially, farther than any other fighter then used in the Pacific. Its long range translated into fighting the Zero over the Japanese home ground.

After some difficulty the P-38 was operating over New Guinea by the end of 1942. In two battles before the new year the 39th Fighter Squadron claimed several aerial victories for two P-38s damaged. Kenney allowed a large number of claims to be confirmed for the sake of morale over statistical accuracy. That he was right in doing so is reflected in the fighting spirit displayed by the P-38 pilots over the next few weeks.

One of the most urgent problems faced by the Allies in New Guinea during the first days of 1943 was the growing strength of Japanese forces coming through the base of Lae on the southern shore of the Huon peninsula. Attacks were made on convoys approaching the area with P-38s acting as dive bombers in addition to engaging Japanese fighters. Between January 6 and January 8, 1943 the 39th Squadron claimed more than thirty Japanese aircraft over Lae while P-40s of the

8th Fighter Squadron accounted for thirteen more.

Without a doubt the P-38 was beginning to make a difference in the total picture of the New Guinea war. It was beginning to be intolerable for Japanese supply lines through Lae and conversely easier for the Allies to reinforce their positions on eastern New Guinea. By the end of January the Japanese were retreating from the Buna area and were more completely confined to the coastal part of the northeast.

The P-38 was not proving to be the best sort of dive-bomber and was relieved of that extra duty to concentrate on intercepting Japanese aircraft. The 9th Fighter Squadron converted to P-38s during January and February giving the Americans two full squadrons to meet the determined Japanese attempt to reinforce Lae by the first of March. The subsequent engagement cost the Japanese four destroyers and a whole convoy of eight transports and became known as the Battle of the Bismarck Sea.

P-38s covering the attack on the convoy claimed thirty Japanese fighters while P-40s accounted for nine more. Actual Japanese losses came to about ten Zeros and probably a similar number of Army Ki-43 "Oscars" for three P-38s and a B-17 shot down. Whatever the numbers it was clear that the P-38 was earning its reputation and justifying Kenney's judgment.

On the back of his spectacular victory in the Bismarck Sea Kenney travelled to Washington to see Generals George Marshall and Hap Arnold for more equipment, especially P-38s. The Pacific had been given secondary priority since Germany with its demonstrated military potential had to be defeated first, but Kenney was offered a compromise: if he could staff it from his own resources, he would get enough P-38s to equip an entire group.

That promised group, plus another group flying another type of fighter, satisfied General Kenney. He rubbed his hands

Vincent Elliott went straight from cadet class 43-D at Williams Field to P-38 transition training and then the 475th in Australia. (Krane)

Major Al Schinz, the first 475th Executive Officer. (Cooper)

Lt. Colonel George Prentice, the first 475th Commander. (Cooper)

and licked his lips before he hurried back to the war to see who in the world he could get to put into his new P-38 unit.

The process went along with amazing swiftness. Kenney sent word down to the squadron level that they were not only obliged to send some of their own meager numbers to the new unit, but that those sent must be the highest quality pilots and ground crewmen. By May of 1943 the cream of Kenney's

Fighter Command was being drawn into the new unit and the howls of protest would be heard for the rest of the war.

Mel Allan was surprised that day in May when he boarded a C-47 for Australia. There were so many men going with him and none that he recognized could be considered substandard. His spirits were buoyed with the thought that maybe this transfer was not going to be so bad, after all.

Known as the "Terrible Four" in the 431st Squadron. Harry Brown, John Hood, Frank Nichols and Bill Haning at Port Moresby. (Cooper)

Chapter II

A GATHERING OF CHAMPIONS

The preface of the official history of the 475th Fighter Group states that, ". . . the urgent necessities of war brought it into existence. It grew up and thrived on combat. Its performance has justified its existence." Perhaps that one statement accurately summarizes the subsequent history of the 475th Fighter Group. It was born of the exigencies of the New Guinea air war, its pilots were the most daring, aggressive and well drilled in the Southwest Pacific and it carved an impressive record from the time that it entered combat.

Even though the Fifth Air Force made stunning use of its equipment on hand, limited numbers of available aircraft severely restricted American airpower in New Guinea. During January 1943, only eighty-two P-38s were on hand while there were no more than about one-hundred P-40s, and just over one hundred and forty P-39s and P-400s.

Against this force there were an estimated three-hundred Japanese fighters within striking distance of areas protected by V Fighter Command. In addition, over one-hundred Japanese bombers were reported on many of those same bases by 23 January. Given the fact that many American fighters were out of service for one reason or another and that others were tied down to non combat duty the Japanese still had the advantage for offensive operations.

Thus, General Kenney's trip to Washington D.C. had some special urgency for him. He hated being on the defensive since his nature was anything but passive. His intention was to strike as quickly and effectively as possible which meant having more combat aircraft than he needed simply to defend targets within range of the Japanese.

Fifth Air Force did receive a fright during April when three separate air raids were directed against its headquarters in the Port Moresby area as well as bases on the eastern tip of New Guinea. On each occasion about one-hundred Japanese raiders were met by less than half that number of Lightnings, Airacobras and Warhawks. Thanks to the determination and growing skill of American fighter pilots each of the attacks was dispersed before serious damage could be inflicted. Japanese records indicate that American interceptors disrupted their formations sufficiently to help dissuade them from further attacks on the Port Moresby area.

However, the vigorous Japanese initiatives strengthened General Kenney's resolve to strike the main enemy bases at Wewak and Rabaul with fighter-escorted daylight bombing raids. One P-47 group was promised to arrive in May, another P-38 group later in the year and the 475th – which was to be equipped with the P-40N – in July.

The Fifth Air Force commander was adamant in his preference for the P-38. He felt that the P-47 was probably a good airplane "in spite of some of the adverse comments from England", but that it did not have the climb rate or range for his needs. The P-40 had served the Fifth Air

475th camp sign with symbols before unit logos were adopted.

Force well, but it had none of the performance required for future plans. P-39 operations had shown a spark of promise before it was decided that the Bell fighter was in the same boat with the P-40.

By the end of the first quarter of 1943 Fifth Air force plans called for one P-47 group in May, another in June and a P-38 group to leave the United States in August 1943. Kenney desired the range and twin-engine safety factor of the P-38 to take the fight to the enemy and became impatient with delays.

Delay did come in the inevitable press of world war, compelling the Fifth Air Force commander to send a message to General George Marshall on May 6, informing him that four-hundred Japanese fighter and bombers were within range of Port Moresby making his need for suitable fighter aircraft had become critical. Marshall replied that P-47s were being

readied to make the 348th Fighter Group prepared for shipment by 12 June 1943.

Another concession that made Kenney especially happy was that he was authorized to activate the 475th Fighter Group as a twin-engine group and the P-38s that Hap Arnold had promised would be available for this unit in June. Kenney was able to write General Arnold late in June: ". . . the P-38s are expected to start arriving any minute now and I think I will meet the boat with a band to welcome them."

However overjoyed he may have been, Kenney still had to man the group from his own resources and that meant squeezing his already meagerly staffed fighter squadrons. While half of the original cadre came from the 11th Airbase Depot and other units outside the Fifth Air Force which sources did supply some eager P-38 pilots who were glad to

Captain John J. Hood on his P-38 "Patsy." (Brown)

"Nick" Nichols, Bill Haning and John Hood on Nichols's command P-38. (Brown)

Grover Gholson already had one Japanese aircraft claimed in a P-39 of the 36th Fighter Squadron when he went into the 432nd Squadron. He also had the experience of being shot down and walking home through the New Guinea jungle before arriving in the new group. (Gholson)

Zach Dean. He once was so incensed by having to abort a mission that he threatened to fight it out with his crewchief, Rozell Stidd. Stidd shrugged off the over-enthusiasm and proudly serviced the P-38 of the ace from the 8th Fighter Group. (Krane Collection)

get into an active combat unit the other half was drained like blood from loudly howling V Fighter Command squadrons.

Not only did they cut to the bone for the sake of the new group, but the other fighter units of the command had to stand for replacement pilots being directed in large numbers to the 475th for the first months of its existence. Throughout its tenure with the Fifth Air Force the 475th was known as Kenney's pet, and more derisively as a "glory-boy outfit."

Perhaps some of the acrimony was understandable in view of the quality of personnel that General Kenney insisted be transferred from other squadrons. Most of the pilots were veterans with at least one aerial victory to their credit and some were proven combat leaders.

One of the pilots who came into the 49th Fighter Group in March 1943 and was quickly bounced into the 431st Fighter Squadron in July represents a case in point. First Lieutenant Thomas Buchanan McGuire was identified as an eager beaver from the first day that he came into the theater. He had already

served a tour in the Aleutians, but from his first days as a fighter pilot in February 1942 he agitated for combat duty.

Oddly enough, considering his later reputation as a non-stop talker, he was considered to be rather taciturn in the 9th Fighter Squadron. His prowess as a P-38 pilot was not unrecognized, but the general consensus was that he would probably get himself killed through his zeal for combat. Something opened him up when he got into the 431st. What seemed to be a quiet young pilot looking for glory turned into the most loquacious pilot still looking for glory. He was talkative to the point of needling his fellow pilots, but no one would ever doubt the courage or skill of this unusual P-38 pilot.

Another pilot destined for glory in the P-38 arrived in HQ, 475th Fighter Group sometime early in October. Major Charles MacDonald had successfully led the P-47 Thunderbolt equipped 340th Fighter Squadron through its first weeks of combat before he jumped at the chance to join the P-38 flying 475th.

Fred Champlin, Tom McGuire, "Pappy" Cline and Frank Monk, later in the war. They were all on hand when the 431st started operations and went on to distinguish themselves.

MacDonald was as introverted as McGuire was now becoming noticeable. Both men earned respect in the 475th. McGuire, grudgingly for his ability as a fighter pilot,and MacDonald both for his piloting skill and for his unassuming competence as a combat leader. In one way, both men would lead this collection of champions: McGuire would be the ranking ace with thirty-eight confirmed victories and MacDonald would inspire a fighting tradition that earned the 475th a reputation for scoring many kills with relatively light losses.

These two pilots would be the top scoring aces of the group, MacDonald trailing McGuire's thirty eight kills with twenty-seven of his own. Captain Danny Roberts followed with fifteen victories, and developed an almost legendary mystique when he took over the 433rd Fighter Squadron in October 1943. Roberts had been a music teacher in Las Vegas, Nevada before the war and had become a fighter pilot in time to join the 80th Fighter Squadron flying P-39 and P-400

Airacobras in mid-1942. He had scored two impressive kills in the Airacobra and two more in the P-38 before he moved to the 475th.

Roberts was unusual as a fighter pilot because of his particularly meticulous habits. He was not given to excessive drinking or smoking and was gentle in speech and manner. Easily winning the trust and affection of his comrades, he showed a knack for leadership that transcended even the remarkable bag of fifteen Japanese aircraft that he was officially credited with shooting down.

Two of the other top aces of the 475th were also in the original cadre. John Loisel had come from the 36th Fighter Squadron and would eventually claim eleven kills. He was noted as a superb pilot and gifted administrator who got involved from the beginning with the 475th's intensive training program. The other pilot was Lieutenant Frank Lent, who started combat as Tom McGuire's wingman and also scored eleven victories. Loisel would become the group's last wartime commander while Lent would unfortunately lose his life in a flying accident over the captured base of Lae after his combat tour was ended.

Mel Allan felt somewhat better when he arrived in Australia sometime in mid-June. He spent a Friday night with his traveling companions at a base that didn't even seem to operate on weekends as if it were peacetime. But after the flight to Brisbane and the first sight of the P-38s on Amberly that the crews were told were theirs, he felt much better. After the New Guinea jungle with its heat and pestilence, constant hard work and occasional danger, the cool climate and proximity of civilization were easy to bear.

There was good chow and plenty of relaxation at first. The inevitable untangling of administrative affairs always meant that the troops on the line had time to explore their new surroundings and pursue favorite pastimes. Allan even had a fifth stripe on his arm as part of the reward for temporarily

leaving the combat zone. When he got to look over some of the P-38s that would soon be in his charge he was happier than he had ever been in his life.

Yet, there was delay caused by problems of personnel and equipment. The official history of the group states that "On 17 June, the first of a contingent from New Guinea, 'shivering in the winter chill of Australia', reported to Amberly Field as cadre personnel for the new organization." Red tape stopped supplies from reaching the group, but the 475th Group was supposed to move to Dobodura by the end of July. Even though there was promised to be an adequate quantity of supplies at the campsite an advanced detail radioed back that there was little in the way of screening, burlap, cement, lumber and piping.

Every working military unit in the field has had to resort to what is known as "midnight requisition" at one time or another. TSgt Orville Joseph, the 475th noncommissioned

Harry Brown, Frank Nichols and Verl Jett. Brown and Nichols came from the 49th Fighter Group and Jett from the 8th Fighter Group. They all had aerial victories to their credit before starting out with the 431st Squadron (Cooper)

Morris L. Sullivan of Moore's Bridge, Alabama had no tech training prior to arriving in Australia on the Queen Mary early in 1942. He was first in the 40th Fighter Squadron, then in the 433rd. Likeable and goodnatured, he advanced from private to master sergeant during his overseas tour. (Ted Hanks)

Identified as TSgt Luke Powell of Junction, Texas. Various pilots flew "The Junction Eagle" from the beginning of 475th operations.

Warren Lewis later on, after he took command of the 433rd Squadron. He was one of the most modest of aces who claimed a number of probable or damaged Japanese aircraft rather than try to get them confirmed. His first victory was scored with the 431st on August 16, 1943. (Cooper)

supply officer, had the knack of browsing through command supply areas armed with bribes for clerks and officers. He managed to liberate some of the material scheduled for the 475th anyway – just a bit sooner than would have been the case through regular channels.

Frank Nichols, commander of the 431st Fighter Squadron, led the first contingent of men to Amberly on June 17 and the rest followed over the next three weeks. Whatever shortages existed on Amberly, most men in the new group were as pleased as Sgt Allan to get fresh meat, eggs, milk that was still liquid and even good liquor. There was also plenty of time to visit nearby centers of population with their high concentrations of females.

Meanwhile, P-38 aircraft began to appear in the 475th line at Amberly. The 432nd Fighter Squadron operational diary records the first contingent of aircraft to reach its engineering section. On 23 June 1943 the first example, a P-38G-5 was taken on charge by 432nd Squadron Engineering. Two P-38H-1s were received on 28 June, three more the next day and another on 30 June.

Five more H-1s were received by engineering before P-38H-l, serial 42-66523 was received on 6 July to be the first P-

David Allen and Art Wenige (right, behind). Allen shot down two bombers with the 7th Fighter Squadron on February 6, 1943. He had four confirmed by August 21 when he claimed two Zeros and an Oscar to become the 475th's fourth ace.

"Every inch a future general." Nichols was a take command kind of leader who often led the 431st with his personal magnetism. (Brown)

Harry Brown was one of the nicest personalities to have around. His pleasant demeanor did not extend to aerial combat, however, and he was as tough as any pilot in the group when it came to handling the enemy.

38 assigned directly to the squadron, becoming the nominal commander's airplane with a large number 140 painted on the nose and tail. By 16 July at least twelve pilots had been assigned to the 432nd and training flights had begun. New personnel were added to the squadron and training continued throughout July. By August 7 preparations were made for the squadron to move north to Dobodura, its first operational base.

Two-hundred and seventeen enlisted men were assigned to the 432nd on July 9. Sixty-three were fresh from the United States while one hundred and fifty four were seasoned troops for the theater. One of the veterans was SSgt George Heap who went into the radio section. He had been assigned to the 8th Fighter Control Squadron early in the war. Volunteering to lead a five man coast-watcher detail to Goodenough Island in 1942, Heap radioed the position of a Japanese landing party in a convoy of barges that was destroyed by air strikes. Heap and his men then went through some harrowing difficulties to escape the ensuing Japanese wrath before reaching safety.

Major Prentice officially took command of the newly formed group at Amberly on June 12, 1943. He started off with the 7th Fighter Squadron, 49th Fighter Group in April of 1941 and went overseas with it in January 1942. In August 1942 he took over the 39th Fighter Squadron until late in March 1943 when he became the commanding officer of the Fighter Training Station at Charters Towers.

One of the most fortunate things to happen to the 475th was the presence of training acumen in the group leadership. It must inevitably turn to the credit of George Prentice that an intensive, if abbreviated, period of training was administered to the combat pilots. The pilots were well grounded in the basic understanding of their P-38 mount and were also ori-

ented to combat tactics that would mean a great deal to the success of the group's initial engagements.

Another edge that Prentice gave his command was the policy of instilling a spirit of the group as a whole. The men were drilled to think of themselves as members of the 475th first and as members of individual units secondly. In actual combat, this policy helped to develop a standard of teamwork that made the 475th difficult to break, no matter what the odds.

During the last days of July 1943 the hurried combat training began to develop a solid cohesiveness in the 475th. By July 30 the 432nd Squadron, for example, had at least thirty pilots on its roster and was flying high-altitude formation training flights, high and low altitude gunnery, ground strafing and single engine drills.

Throughout the training period it was necessary to test fly the P-38s as they arrived as well as organize the command structure of the squadrons. Captain Tomkins became the commanding officer of the 432nd with Captain Danny Roberts as Ops Officer. Captain Ronald Malloch was the squadron Intelligence Officer and the enlisted cadre was headed up by First Sergeant Ernest A. Barnes.

By the time Captain Nichols was ready to take the 431st Fighter Squadron to New Guinea he had Captain John Hood as Operations Officer and Captains Harry Brown and William Haning as flight commanders. Both of these latter officers were from the 9th Fighter Squadron. The third flight commander was Captain Verl Jett. Lieutenants Warren Lewis, Marion Kirby, Thomas McGuire and Lowell Lutton were flight leaders.

In the group structure, Captain Albert Schinz assumed his position as initial executive officer while Captain Meryl Smith was Ops Officer. Captain Claude Stubbs became group

S-4 and Captain LeRoy H. Richardson came from Hq, V Fighter Command to assume duty as group adjutant. Group Intelligence Officer was Captain Bennett Oliver. Finally, the Australian liaison officer was Captain Frederick D. Percival from the 1st Australian Air Group.

Training moved along rapidly and, by the time they were ready to move north, some of the 475th pilots were able to fly their P-38s directly to 12-mile drome and Jackson strip near Port Moresby. During the attempt to fly these fighters to New Guinea at least one P-38 crashed on takeoff, but the pilot, Lt. John A. Cohn, escaped with minor burns. Other group personnel arrived by C-47 transports or landed at Dobodura on the S.S. Joseph Lane.

Bill Haning (above) took his various jobs seriously, perhaps too seriously according to Dennis Cooper (below), and suffered more than just the weight of combat responsibility. (Cooper)

Captain Campbell P.M. Wilson. He came from the 8th Fighter Group and rose to command the 433rd Squadron.

Johnny Loisel. The soft-spoken ace from Nebraska managed difficult tasks with tacit aplomb. He was responsible for much of the orientation of pilots and crews to their P-38 aircraft.

The legendary Danny Roberts. Somehow, he was able to project his personality and inspire confidence with it. He picked up the spirit of the 433rd and elevated it from a good squadron to an inspired one. (Cooper)

Clarence Wilmarth was a friend and classmate of Danny Roberts. He was rated in the same category of warm human being as Roberts by Ted Hanks. (Hanks)

Clarence Rieman later in the war at Hollandia. He was typical of the superior sort of pilot who was drafted from other squadrons. By the time he went home, he had participated in most major engagements of the 433rd and claimed two Japanese aircraft. (Rieman)

Joe McKeon is now a medical doctor in California. He had one victory with the 35th Squadron and scored four more with the 433rd. (Cooper)

Art Wenige (pronounced Wen'a - gee) claimed five victories with the 475th, the last being shot down over Rabaul in November 1943. (Cooper)

Young Fred Champlin would become a standard of the 431st throughout 1943 and 1944. He is also using Hood's P-38 to pose for his picture.

Bob "Pappy" Cline on Hood's P-38. Cline was one of the best leaders in the 475th and was reputed to be an outstanding fighter pilot.

Franklin Wilson Ekdahl at the same awards ceremony pictured in the preceding chapter, facing General Wurtsmith. Ekdahl was also one of the standards of the 431st.

433rd Squadron #190 on a maintenance stand early in the history of the group. (Krane collection)

John Ehlinger was one of the six 39th Fighter Squadron pilots who formed the nucleus of the 433rd original cadre.

Verl Jett in the Hawaiian Islands before he was involved in the whirlwind of combat with the 8th and then the 475th Groups. (Brown)

Henry Condon later in the war. He went into the 432nd as a Flight Officer and commanded the squadron until his death in January 1945. (Krane)

Paul Morriss also came from the 8th Fighter Group and ended up in the 431st Squadron. He was another pilot who took combat seriously enough to affect him deeply, according to Cooper. His face shows his attitude in this detail from a briefing photo for the first 431st mission on August 12, 1943. (Cooper)

Sign well-known to travellers from Port Moresby to the American camps. The other side proclaimed, "I told you so!" (Cooper)

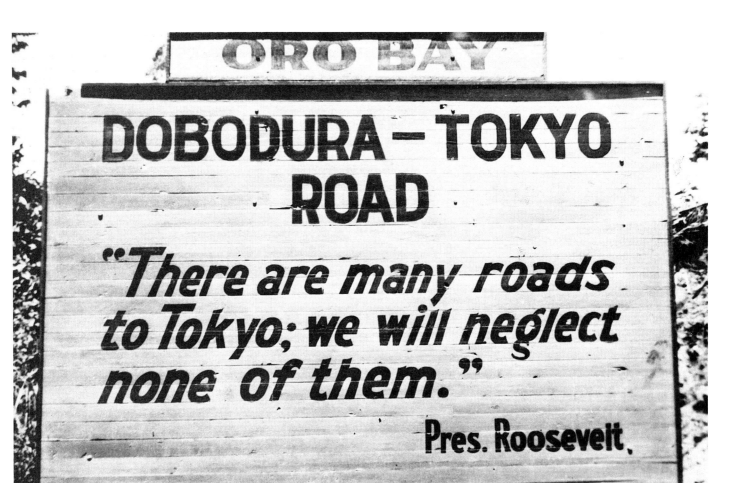

Sign leading traffic to Dobodura road.

Chapter III

DOBODURA & COMBAT

Captain Richardson led the water echelon aboard a liberty ship on August 5 toward Oro Bay and eventually Dobodura some miles to the northwest of Oro Bay. The ground echelon arrived at Dobodura on August 14. By the end of the day on August 7 the air echelon had moved into the Port Moresby fields at the campsite formerly occupied by the 35th Fighter Control Squadron.

When the air echelons finally regrouped in the next few days the 431st Squadron was located at Twelve-mile strip; the 432nd at Ward's Drome and the 433rd at Jackson's Drome. The men of the 475th who not there before were not impressed by their first sight of the Port Moresby area. Even those who had flown from the area before were reminded why they were glad to get back to Australia, if only for a few weeks.

Seven-mile, where the 433rd was stationed for the moment, was typical of the bases. The base was named Jackson's Drome for the commander of the famed 75 RAAF Squadron who was killed in the early defense of Port Moresby. Only ninety-six feet above sea level, the heat was oppressive on the base in the afternoon. Any rare breeze happening along simply stirred up a choking dust. Flies buzzed constantly and water was available only through the boiling Lister bags that hung on any convenient post.

Several patrol missions were flown directly from Port Moresby by 475th formations beginning on August 12, 1943. Pilots of the group were still gaining proficiency in the P-38 and learning the basic combat tactics that would make the 475th victorious in the days to come. The most important lesson stressed to the pilots was the integrity of the battle formation. Later in the war, captured Japanese airmen would testify that the P-38 was a most fearful adversary when it maintained formation attacks.

The 475th ground and water echelons worked quickly to prepare the Dobodura strips at Horanda to receive the P-38s and their pilots. Dobodura differed only slightly from Port Moresby. Heavy timber surrounded the flat ground of the area with dense marshland bordering the Giruo and Samboga rivers that virtually converged off the end of Horanda Strip West number 4. Since the ground was slightly moister than at Port Moresby, there was a bit more breeze and less dust in the faces of groundcrewmen.

On August 16, 1943 P-38s of the 475th flew their seventh combat mission. 433rd P-38s had already flown to Dobodura and were ordered back on the 16th. During the afternoon the 431st Squadron was off on a mission to escort transports to the base at Tsili Tsili near Marilinan in the highlands west of the Huon Gulf.

Captain Harry Brown led fifteen P-38s of the 431st off the ground just a few minutes before 12:30 in the afternoon. Somewhere along the way one of the P-38s had to abort because of oxygen

475th theater at center of Dobodura campsite (Hanks)

Papuan natives building grass-shack at Dobodura. Perhaps famed Club 38? (Hanks)

failure and another turned back due to some other sort of mechanical problem. Brown was left with twelve P-38s behind him to protect the transports, but it should be a routine escort.

The day before, over the same territory that the 431st was now taking the P-47s, a formation of 35th Fighter Group fighters had given the P-39 its best day in aerial combat over New Guinea. The 40th and 41st Squadrons had intercepted an incoming raid of Ki-48 "Lily" bombers escorted by Ki-43 "Oscar" fighters. Fourteen Japanese aircraft had been claimed destroyed for the loss of four Airacobras.

In spite of the report of the previous day's combat there was not a great deal of anticipation of impending contact with the enemy. The P-38s were flying at 20,000 feet through cumulus clouds that dispersed into clear sky over Tsili Tsili; it was peaceful in the sky over the landing area at fifteen minutes past one in the afternoon.

Lieutenant David Allen, element leader of Green Flight, first saw the enemy a lone single-engine fighter – off to the west and called in the alert. His flight was the last of the high cover circling north of the landing transports when the entire escort reversed course and headed in the direction of the sighted enemy.

475th landing at Dobodura (Lenard via Krane Collection)

Awards ceremony by 433rd Squadron. 175 was the usual mount of Charles Grice, formerly of the 39th Fighter Squadron, which explains the white shark's teeth, usually a talisman on 39th FS P-38s. (Lenard via Krane collection)

433rd line at Dobodura. (Lenard via Krane Collection)

Lieutenant Allen was becoming a little eager when the flights ahead made excited shouts over radio about enemy aircraft coming down from above. Allen could see the P-38s of the first two flights being engaged by single-engine fighters. He could even see the lead P-38 undoubtedly Captain Brown's firing at what appeared to be a Zero or Oscar that was going down in smoke and flame.

Within a few seconds it was Allen's turn when he flew into the middle of the fight. The first few shots that he made were wasted before he settled down and set his sights on an Oscar that was turning in to attack his flight leader. Allen was impressed by the firepower of the P-38 when an accurate burst of the concentrated four fifty calibers and single twenty millimeter cannon ripped into the wingroot of the Oscar. The entire wing tore off and the Oscar spun into the jungle below like a hellish autumn leaf.

Allen quickly recovered in time to fire at another Oscar that was coming at him. He decided to let that one go since one of his squadronmates roared by in another P-38, hot on the tail of the Japanese. Over Marilinan Allen encountered another Japanese that he identified as a B5n "Kate" torpedo bomber and shot it down in flames northwest of the Marilinan airfield. Harry Brown managed to witness both of Allen's victories.

For his part, Brown was listening to warnings over the radio by the controller of enemy plots in the area when Allen's call prompted him to turn around. Eight Japanese fighters were coming down and were almost in firing position when Brown led his flight into a dive to escape. Gaining speed in the dive he zoomed back to altitude into the sun for a favorable position.

For just a brief second one of the Oscars presented itself as a greenish/brown target, dancing in the sparkling sunlight.

Brown missed with the one shot he got off, but pulled up in time to see another Oscar dead ahead and slightly above. This time the Japanese was surprised and took hits until flame erupted from the right side of the cockpit and the Oscar went down into the jungle.

The action was furious now and Brown turned to the right to fire at another hapless Oscar that caught fire after receiving accurate gunbursts from Brown and at least one other P-38. Another Oscar pulled almost straight up into a hammerhead stall after attacking a P-38, but lingered long enough for Brown to explode it with a single burst of fire.

Lieutenant Lowell Lutton had been a classmate of the top American ace of the war, Dick Bong, at the Rankin Air Academy in 1942. He was good enough himself to become a flight leader in the 431st and was leading Green Flight in this combat. When his element leader, Lieutenant Allen, called out the single enemy fighter he watched Brown's flight dive away and the second flight turn into the Japanese attack. Lutton turned his flight into a climbing attack through the Japanese attackers, firing all the way.

When he dived away after the Oscars had passed, Lutton could see that only his wingman, Lieutenant Orville Blythe, was still with him, but decided to continue the battle anyway. In and out of the pattern of combat the two P-38s skidded and rolled, firing snap deflection shots until two Oscars came in for a head on pass. The first turned away before getting into range of Lutton's guns but the second stayed on its course, the two heavy guns above its engine flashing a staccato of dirty orange lights well out of range.

Lutton opened up with his own guns when the Oscar was in his sights and scored a stunning hit that tore away part of the engine and canopy. The Oscar went straight down from

The guns on Mankin's P-38 jammed at that time forcing him and Smith out of the fight. However, it was mainly due to the fact that they stayed together throughout the fight that three enemy aircraft were destroyed by them without much damage in return.

It was much the same for Lieutenant Art Wenige who shot down one Oscar unfortunate enough to cross his line of flight and another that was on the tail of a P-38 in his flight. Lieutenant Warren Lewis declined to claim an Oscar as destroyed that he shot off the tail of a P-47 operating in the same area, but subsequent testimony from ground personnel at Tsili Tsili confirmed that the aircraft had crashed near their site. It gave the unusually modest Warren Lewis his first confirmed victory.

Since they had been in contact with the enemy for almost half an hour and the Japanese fighters were now well scattered, Brown called his fighters to rendezvous over Wau. Twelve of the P-38s returned immediately to Port Moresby, Lieutenant Blythe's aircraft being temporarily detained at Marilinin. His P-38 was the worst American casualty of the battle with only one other P-38 suffering damage at all with hits in an intercooler.

Japanese Army Air Force records are incomplete for this engagement, but even if the 431st overclaimed significantly the victory was an impressive one. The P-38s stayed in pairs and formed a mutually protective pattern for the individual flights. While one flight dived another turned into the Japanese and the third climbed for advantage. The result was devastating, even if the American claims are cut to a fraction.

The big push by the Allies to neutralize Wewak began later that same night with heavy attacks by B-24s. Both Wewak itself, and But airdrome were heavily hit. The next day another massive blow was scheduled using B-25s and P-38s at low level.

Maintenance and armory on a 433rd P-38 at Dobodura.

15,000 feet almost directly over the Marilinan airfield. Blythe watched the enemy fighter go down until another Oscar behind him started scoring hits that caused smoke to pour from his right engine.

He and Lutton had to dive away from the attacker at his tail. When the air was clear Lutton found that he could not contact Blythe, but could see that his propeller was feathered and that his P-38 was too badly shot up to maintain altitude. Lutton stayed with him until he was able to land at Marilinan and then flew back to Port Moresby.

Jack Mankin and his wingman, Lieutenant Paul Smith, had dived down to help out some of the P-38s under attack after dropping their belly tanks. Mankin got to within one-hundred yards of an Oscar and sent it down in flames. He and Smith then attacked another Oscar flying along with what appeared to be a Kate. The two Japanese pulled up and the two P-38s fired in unison setting the two targets afire at almost the same precise moment.

Jack Fisk and Clarence J. Rienan next to Fisk's 433rd P-38 #191, probably August or September 1943.

433rd line at Dobodura on an overcast day.

This raid was one of the most successful by any Allied force in New Guinea. About one-hundred Japanese aircraft were put out of action on the ground, including almost every Ki-61 "Tony" fighter of the 78th Sentai (air regiment). The only Tony unit operating in New Guinea for about a month afterward was the 68th Sentai.

While the surprise attack of August 17 had devastated Japanese aircraft and equipment, the campaign to crush the base went on. The next day forty-eight P-38s of the 475th were prepared to escort more than fifty B-25 strafers and a group of high-flying B-24s. By seven o'clock in the morning the 475th was off to do battle for the first time as a group.

The Japanese had concentrated its 4th Air Army in the Wewak area only days before the Fifth Air force began pounding the base. On this morning of August 18, there were various elements of four Oscar units and only six Tonys of the 68th Sentai available to counter the Americans. There was little in the way of advance warning after the confusion of damage the day before. Some Oscars of the 24th Sentai were on patrol and reported an enemy force of about forty aircraft over Hansa Bay, headed north.

There was an immediate alert and every available Japanese fighter was scrambled. Seven 24th Sentai Oscars, five 68th Sentai Tonys and two 13th Sentai Ki-45 twin-engine

The objective of 475th training was to have every mission go out and come back like this 433rd formation. (Hanks)

Col. Prentice on the line with Major Schinz, a Red Cross girl, Bill Haning, Major Nichols and Lt. Phillips in the background. The Red Cross girl is just serving refreshments to a group of returned pilots. (Cooper)

On the strip: alert shack at Dobodura. Left, foreground: Veit, behind him Monk, writing is Lent, laughing is Allen, sleeping is unknown; background laughing is Kirby, background far right is Morris, foreground right Samms then Houseworth. (Cooper)

"Nicks" took off from Boram plus nine 59th Sentai Oscars from But airdrome to intercept the American raid. Japanese forces were certainly crippled, but not unwilling to meet the enemy in the air.

At a higher altitude than the raiding B-25s the B-24s got good bombing results in spite of the murky weather that seemed to cover the ground in a watery film. P-38s were able to protect the heavies from relatively weak interception, but the B-25 strafers ran into a hornet's nest while they raced in to shoot up about sixty aircraft on the ground (Japanese records state that only about twenty-eight were serviceable).

Nanba Shigeki was a Chutai (approximately flight) leader of the 59th Sentai and led five other Oscars into an attack on the B-25s when he noticed what seemed to be dozens of P-38s

Verl Jett's first P-38. On the right side of the gun bay is a rendition of "Fifinella", the Women's Service Pilots mascot.

Champlin's P-38 in its element over Dobodura. (Krane collection)

coming down from the mist above. The 431st Fighter Squadron had been covering the B-25s at an altitude of 4,500 feet and were slamming into the enemy interceptors to rescue the American strafers.

Shigeki was in a disadvantageous position, but bravely continued fighting until he noticed that about ten P-38s had latched onto his tail. He discovered to his chagrin that the Oscar would not outclimb the American P-38s and the first of the P-38s was in firing range. The Oscar was hit and went into a wild dive that the pilot was able to control long enough to crashland on But West field. It is possible that the Oscar was so skillfully maneuvered that as many as four P-38 pilots claimed it shot down before it actually landed. Two other 59th Sentai pilots were killed in the battle.

Major Albert Schinz had come down from group headquarters to lead the 431st on low escort. When the strafers went in on their run he saw the Japanese interceptors coming

straight down to the attack and he immediately switched gas tanks and dropped his external fuel.

He shoved controls forward to increase speed and rpm just as six enemy aircraft jumped his flight from above. Schinz was forced to turn away from defending the B-25s to confront the immediate danger. Firing quickly, he got a shot at the lead Oscar before the two planes passed at terrific speed. Then he wheeled around and got on the tail of another Oscar for more measured aim while the Japanese pilot frantically dodged in and out of the low clouds.

The Japanese fighter took some deadly hits at close range from Schinz's guns and crashed into the sea off Dagua strip. Schinz was low enough now to see the B-25s still strafing with Oscars boring right in after them. He chased one Oscar off the tail of a B-25 just before the bombers finished their run and pulled up into the overcast.

Meanwhile, Captain Verl Jett had led his White Flight down into a good bounce on the Oscars attacking the B-25s. Jett and his wingman, Lieutenant Bob "Pappy" Cline, got into the middle of the fight and Jett shot the wing off one Oscar on his first pass and set fire to another Japanese fighter on his second. Cline dutifully followed Jett and flamed a third Oscar that was attacking his flight leader.

Lieutenant Ed Czarnecki had led Jett's second element down through the mist onto the backs of the Oscars snapping at the heels of the B-25s. He fired at one Oscar and saw red flashes of fire on the enemy wing before breaking away and attacking another that had outmaneuvered a P-38.

This Oscar was a set-up for a tail shot from Czarnecki who shot at his target until it pulled up with black smoke and flame coming from its nose. Another Oscar got out in front of the P-38's guns and Czarnecki fired until he saw the right wing explode into an orange mass of flames. During a subsequent attack on yet another Oscar Czarnecki's guns became hopelessly jammed and he had to leave the combat area.

Camp area in Dobodura.

Dobodura from the air. (Krane collection)

431st Squadron ready for first mission briefing; August 12,1943. L-R: (Probably) Australian Liaison Frederick Percival, Captain Haning conducting the briefing while pilots soberly absorb data for a mission that proved routine. (Cooper)

Lowell Lutton led Blue Flight down after Schinz and Jett had made their first passes. He fired at a climbing Oscar that managed to maneuver away. When he turned his P-38 back in the direction of the combat he saw an enemy fighter falling through the sky with flames coming from its engine; his wingman, Lieutenant Donno Bellows, had shot it down.

His flight disappeared behind him before Lutton knew what was happening, so he quickly tacked onto two P-38s flown by Lieutenants Tom McGuire and Frank Lent of the last flight. McGuire's number four man had aborted before reaching the combat area and his element leader, Lieutenant Bob Sieber, became separated in the wild maneuvering.

An Oscar had maneuvered onto the tail of Lent's P-38. He called for help and dived, skidding to the right to keep from being killed by the enemy close on his tail. Apparently, both Lutton and McGuire shot at this plane since they both claimed

hitting a Japanese fighter that exploded around the cockpit at this time.

Lent had the opportunity to return the favor shortly thereafter when he shot off part of the canopy, engine cowling and tail section of an Oscar that had rolled in on McGuire's tail. Both Lutton and McGuire chased an Oscar down to about 1,000 feet off the water of the bay and Lent came along in time see the enemy fighter crash into the water.

The action was at its zenith and Lent was able to witness three single-engine fighters falling at the same time, one in flames and the other two streaming heavy black smoke. He also noticed several large fires on the beach and runway and a large, dark barge offshore.

Heavy flak covered the target area when Lent and McGuire left to cover three retreating B-25s of the 405th Bomb Squadron. About ten minutes out of the weathered-in Wewak area

432nd dispersed P-38s. Campbell Wilson usually flew #145 in the middle ground. (Krane collection)

More 432nd P-38s at Dobodura. (Krane collection)

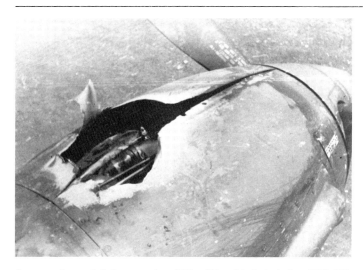

Damage that ended the operational life of Tom McGuire's first PUDGY on August 29, 1943. McGuire's P-38 was hit by heavy caliber fire after he scored his seventh victory. (Alder via Krane collection.)

the clouds began to abate somewhat with the promise of clearer skies toward the Sepik River ahead.

Unfortunately, the increased visibility also made things easier for the determined Japanese interceptors and a single Tony made a pass at the bombers. McGuire turned into the Japanese fighter and fired two or three bursts that started it smoking. Gunners on the B-25s and Lent all saw the Tony crash below.

McGuire and Lent escorted the bombers to Bena-Bena then refueled at Marilinan before returning home. Only one 431st P-38 did not return; Lieutenant Ralph Schimdt was last seen leaving his flight during the battle. Captain Bill Waldman claimed a Zero near But drome for the 432nd Squadron's first confirmed victory, and that unit's only casualty was a P-38 that landed at 30-mile drome with a rough engine.

Japanese sources contradict the extensive American claims and limit the losses to two 59th Sentai pilots killed and another Oscar severely damaged.* There is little doubt that McGuire got his Tony, but the Japanese only report one 68th machine crash-landing.

August 19 was a day of brief respite for the group. The news was that the camps at Dobodura were shaping up niccly and would be ready by the end of the month. Life at the Port Moresby dromes was anything but pleasant and the prospect of moving forward to more comfortable emplacements was cheering.

Another Wewak mission was flown on the next day without much more action than chasing off several unwilling Japanese fighters. Some of the 432nd Squadron stopped to

refuel at Marilinan and most of them were off again within a few hours. Lieutenant Allan Camp was one of the last off.

He must have become confused and lost his way, because he was next heard of trying to land at Karema about fifty miles northwest of Port Moresby. Apparently his disorientation caused him to overshoot the strip and crash into the Karema river inlet near the Terapo Mission. His loss was the first for the 432nd.

Nevertheless, the fighting spirit of the group was demonstrated again the next day and the 432nd was aggressively involved in the action. 432nd Squadron flights were led by Danny Roberts, Lieutenant Fred "Squareloop" Harris, Noel Lundy and Captain Waldman, who had scored the Squadron's first victory just three days before.

Leading one of the elements in Harris's flight was 1st Lieutenant John Loisel with Lieutenant Paul Lucas as his wingman. Loisel was an unassuming young Nebraskan, tall, lean and fair with a quiet nature that belied his energy in passing on his expertise to the rest of the 475th. He studied the P-38 with a scholar's zeal and taught his fellow pilots to trust and love the airplane as he came to regard it.

When Clover leader (Clover was the 432nd's call sign and Danny Roberts was leading this mission) called out to the squadron to drop tanks near Wewak, Loisel waited until he saw the glistening ovals flip away from the lead flight before he got rid of his own. He and Lucas then jumped headlong into the fight and made successive passes at a Tony, a pair of single-engine Japanese fighters, and a twin-engine fighter before settling on an Oscar that was in a slight bank to the right.

The Oscar was able to turn into Loisel's line of flight and the two enemy fighters were able to blaze away at each other briefly before the Japanese ducked under the P-38's left wing. Lucas had a bit more time and aimed a good burst before the Oscar passed under his P-38, too. Lucas could see the flash of yellow flame reflecting off the underside of his own P-38 while Loisel witnessed the Oscar actually burning and falling to the jungle below.

Loisel then jumped a Tony that showed some fight by turning around into a head on pass. Again Loisel pressed the attack very close and saw large pieces of the Tony's wing flying off. Lucas stayed right with Loisel and saw the Tony heading down with most of one wing shredded off. He learned later that his diligence almost had its cost when a large hole was discovered in one of his own wings, certainly made by a detached part of the Tony.

Another Tony in the same local engagement south of But and Dagua took an even deadlier dose from Loisel's guns. The Japanese fighter staggered and began to burn and smoke at low altitude. Lucas saw the Tony turn over on its back and

* NOTE: Shigeki was again attacked by P-38s on this mission. Two of the Americans apparently thought they had finished him off over the sea and turned away, allowing the Oscar to attack them and perhaps be identified as an entirely different machine.

#163 was usually flown by Lt. Jim Farris of the 432nd Squadron. The ship at right is probably the McDui, a beached landmark at Dobodura. (Australian War Memoriaa)

Tom McGuire being awarded one of his medals by General Wurtsmith.

plunge straight down below a cloud layer at about 1,500 feet. Both American pilots were sure that the doomed plane smacked into a hill just under the cloud.

Danny Roberts and his wingman, John Michener, had plunged into the initial combat with Japanese interceptors over Boram, about ten minutes from Wewak, when calls came from the B-25s that "Zeros (were) over target." Roberts and Michener were alone at about 6,000 feet, but decided to go to Wewak while the rest of the 432nd was heavily engaged. Shortly, the two P-38s arrived and the Americans were hardly encouraged to find twenty to twenty-five Japanese fighters visible in the sky.

Six Oscars on the tail of a single P-38 caused an immediate response by Roberts to rush in for a rescue attempt. Unfortunately, the P-38 under attack was already badly hit and in trouble, falling in flames before it could be saved by the intervention of Roberts and Michener (three 80th Squadron P-38s were lost in this same battle for one of that unit's worst losses).

Roberts exacted quick retaliation when he picked out the second Oscar in the string and got some good hits in the engine. The Oscar pilot was obviously surprised because he turned away from the attack and exposed the glistening belly of his fighter long enough for a killing shot and went in smoking near But airdrome. Another Oscar that Roberts had damaged tried to climb away from the fight, but only managed to fall into the sea off Dagua strip.

Harry Brown, became an ace on August 16, 1943 during the first 475th aerial engagement. (Cooper) Right: Brown's P-38 which Frank Monk inherited at the end of 1943.

One of the Ki-45 twin-engine "Nick" fighters of the 13th Sentai got off on the interception with two recently acquired Oscars, one of which was flown by 2nd Chutai commander, Lt. Harada Ryoshei. Apparently this trio of fighters ran into a host of P-38s and was wiped out.

2nd Lieutenant Billy Gresham had lost his leader, Lieutenant Grover Gholson, when the two of them attacked a pair of Tonys. While Gresham was looking around to find Gholson he saw three P-38s attacking a single Nick. The Nick was hit and trailing smoke, but, amazingly, still managed to roll and dive out of the way of the American guns.

Seeing an opportunity, Gresham got on the tail of the Nick when it was momentarily in the clear and fired at about 15 degree deflection from behind until the P-38 ran out of ammunition. The Nick continued its steep, flaming glide until it struck the ground. It is possible that this was the single 13th Sentai Nick that got off the ground and was claimed by at least four P-38s.

One of the pilots who may have attacked this same twin-engine Japanese fighter was Tom McGuire who claimed to have damaged one that outturned him. His faithful wingman, Frank Lent, was close enough to take the Nick head on and claimed that he had shot it down. Campbell Wilson's combat report also closely tallies with McGuire's and he also may have claimed the same Nick.

McGuire did subsequently claim a "Zeke" that Lent saw crash in flames from low altitude, and another that was seen to go down by another 431st Squadron pilot. When these two kills were confirmed McGuire became the first pilot to score five victories entirely in the 475th Fighter Group.

As a whole, the 431st Squadron scored eleven victories. Three Oscars were credited to Lieutenant David Allen to make him an ace of the group, also. Captain Jett was credited with a Nick and Frank Nichols got an Oscar for his fifth

victory. 2nd Lt Ronald Dunlap got another fighter and Ed Czarnecki claimed a fixed landing gear fighter known as the Ki-27 "Nate", which in truth must have been another Oscar whose landing gear had simply been extended in flight.

Fred Harris and Elliott Summer had become separated from the rest of their flight in the wild fighting at the beginning of the engagement and had a time of it. Summer watched Harris shoot one Tony into flaming pieces and then sent an Oscar down in flames himself. Harris damaged four other Tonys before he settled down on one that had just shot out an engine on a P-38. Summer was there again to witness Harris set the Japanese plane on fire with his first burst.

Twenty-three Japanese aircraft would be added to the tally of the new group after this rough mission. In three combats since it began operations the 475th had claimed about fifty Japanese aircraft shot down for the loss of one P-38 and pilot in battle and another in an operational accident. Even if only a fraction of that total were, in fact, true, it was still a remarkable debut for any combat unit.

Bill O'Brien, "Pappy" Cline and Fred Champlin before a typical 475th dwelling at Dobodura in August 1943. (via Jeff Ethell)

Informal portrait of (mostly) 431st Squadron at Port Moresby: Standing (l to r): O'Brien, Knox, Morriss, Samms, Champlin, Elliott, McBreen, Cline Czarnecki, Bellows, Dunlap, Kirby, Brown. Kneeling (l to r): Smith, Allen, Wenige, Phillips, Hood, Blythe, Jett, Hedrick, Lewis, Gronemeyer, Houseworth, Lent, Cohn.

The 431st Squadron* rounded out its engagements for the month of August with a medium altitude (17,000 feet) escort of B-24s back to Wewak on August 29. Sixteen P-38s covered the bombers until a number of Oscars came down on the bombers and 1st Lieutenant Harold Holze got one of them for his first air victory.

Tom McGuire was leading Blue Flight when he saw two Japanese single-engine fighters make a pass at the bombers about a thousand feet below. Blue Flight had only three P-38s since McGuire's wingman had aborted some time ago, but McGuire immediately ordered tanks dropped and led the remaining element to the attack.

Blue Flight was approaching the two Japanese Oscars quickly when their pilots noticed the danger and dived to the left. For some reason, the lead Oscar pulled up and exposed its silvery fuselage long enough for McGuire to fire and send it down in flames.

By now McGuire looked around and noticed that he was alone, but the sight of a Tony heading in for the bombers quickly drew him in that direction. He faced the Tony head on and turned around in time to get several good bursts from rear deflection that set the Japanese fighter spinning down in flames.

Heading up for some more altitude, McGuire noticed tracers passing his P-38 and glanced around to find three radial-engine fighters and another Tony coming down on his tail.* He dived like an alarmed fish for the safety of the depths below, but one of the pursuers was a good enough shot to hit the P-38's left engine with an explosive shell and set it on fire

* NOTE: The original call sign for the 431st was HADES, which was changed to DADDY during November 1944.

* NOTE: McGuire was probably one of the four P-38s claimed by the Japanese during this engagement. No other P-38s on the mission were as heavily damaged as McGuire's. In turn, the 59th Sentai lost four Oscars, most likely including McGuire's claim.

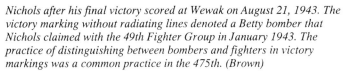

Nichols after his final victory scored at Wewak on August 21, 1943. The victory marking without radiating lines denoted a Betty bomber that Nichols claimed with the 49th Fighter Group in January 1943. The practice of distinguishing between bombers and fighters in victory markings was a common practice in the 475th. (Brown)

Aerial view of Wewak with the airfield at right. (Krane collection)

It was a moment for decisive action and McGuire's life depended on making the right decision to put out the fire while evading the enemy now breathing down his neck. He shut off the burning engine and slipped the P-38 down to 9,000 feet where the fire went out. He continued the dive and lost three of his attackers before the other gave up when the P-38 entered a merciful cloudbank at 4,000 feet.

McGuire made a one-engine landing at Marilinan. His Oscar claim was confirmed by some of the crews of the 320th Bomb Squadron and his remarkable calm that had saved his life only a few hours before now convinced intelligence officers of the 41st Fighter Squadron that he had also shot down the Tony. He had seven confirmed kills in three battles, but his first P-38 was a total loss.

Ssgt. Heap and unidentified pal, probably at Dobudura. (Heap via Maxwell)

Chapter IV

HUON GULF –
SEPTEMBER 1943

O n August 31, 1943 the first detachment of nineteen 432nd Squadron P-38s flew from Port Moresby to Dobodura. Captain Danny Roberts led the first flight which included Campbell Wilson, Jim Farris and Christopher Bartlett. "Squareloop" Harris led the second flight with Captain Waldman heading the third and Lieutenant John Loisel bringing up the rear with a last flight of three.

The 433rd Squadron moved into its new camp area in Dobodura on September 1 after cutting down trees to form tent frames from the surrounding forest. An escort to Cape Gloucester was flown by the hustling 433rd during the same day. The squadron put on a display of aerobatics on returning from the mission as a form of thanks to the 43rd Engineers for completing the strip.

Another mission flown on the same day by the 432nd had a somewhat different purpose when eight P-38s led by Captain Donald Byars escorted the "Bully Beef Bombers", or food ration laden C-47 transports, into Dobodura. Another flight of Bully Beef Bombers was escorted in by a flight led by Captain Waldman later in the day. Most Americans thoroughly hated the unpleasant Australian bully beef; some of the pilots jokingly asked if anybody would actually stop a Zero from shooting down any transport carrying the stuff.

Happy Crews and Unhappy Crews

Brand-new Technical Sergeant Melvin J. Allan was much happier with his assignment to the 431st Fighter Squadron and with the five stripes he now sported on his arm. He liked working on the P-38 and the accommodations at Dobodura were nominally a bit drier and less stagnant than the inland quarters of the Port Moresby strips.

One of the small comforts that began to appear during the early days at Dobodura was known as the "Smoko", or simple repast available to crews on the line. Usually the smoko was prepared over a small gasoline or electric stove. Originally it was something no more complicated than coffee, but by the time the 475th arrived in the Philippines it included fried chicken, eggs or even carabao steaks.

None of these comforts would have cheered the bleak spirits of the 433rd groundcrews, however. Major Low's policy of withholding promotions for higher efficiency was having a disastrous effect on morale. The crews were all highly motivated to begin with, and most of them were switching from single-engine to twin-engine maintenance which meant an almost automatic promotion. The bitter pill was made even more unpalatable when the other two squadrons made enlisted promotions on schedule.

SSgt. Ted Hanks had mixed feelings about his lot with the 475th Fighter Group. He loved working with airplanes and became especially fond of the P-38, considering it the finest piece of equipment he had worked on up until that time. In addition, he had been requested as crew chief by Captain Clarence Wilmarth of his old 40th Fighter Squadron. Wilmarth had graduated from the same aviation cadet class as Danny

Roberts (41-G) and Hanks placed them in the same category as competent officers and decent, likeable human beings.

What Hanks did not like about his squadron was primarily the original first sergeant who managed to make life unbearable for more than one enlisted man in the 433rd. Hanks thought him a completely unreasonable ogre who delighted in ruling the men under his authority with peremptory disdain. More than one enlisted man in the squadron ultimately blamed the squadron commander, Major Martin Low, for the appointment of the first sergeant as well as other unfortunate decisions in the formation of the 433rd.

The truth of the matter is as was stated before; Major Low was a senior pilot in the 40th Fighter Squadron who in one sense was thrust into the command of the new squadron without much choice or real preparation, He probably had around 150 combat missions to his credit by the time he was sent home with other worn down veteran pilots late in 1943. Of course, he did volunteer for another tour in the European theater and commanded another squadron, perhaps learning a bit about labor relations with enlisted men in the process.

Whatever the friction may have been between the administration and ground crews of the 433rd there was no denying that the squadron was beginning to strain at the bit to get into

Below: Lt. Donno Bellows's #130 "Piss Pot Pete", "Linda Kay" on port engine cowling (above).

Lunch on the line for 431st Squadron pilots: far side of table (l to r); O'Brien, Champlin, Cline, Kidd, Lent, Donald, Cortner. Near side: Ekdahl, Elliott and Morriss. (Cooper)

431st Engineering Section at Dobodura.

combat. So far, the most tangible bit of action the P-38s of the 433rd had seen was on the August 18 mission to Wewak when Captain Herbert Jordan got close enough to fire at an Oscar with no visible results observed.

First Blood for the 433rd

The pilots of the 433rd knew that they would be off on a mission the next day; September 2, 1943. They are keen to be off, so before the sun is even barely beginning to show its rays over the treetops they are in the operations room waiting for instructions. The target and rendezvous are still unknown, but the feeling is that an engagement with Japanese aircraft is imminent.

One of the young pilots waiting in the Ops shack is Lieutenant Calvin "Bud" Wire, a strapping, black-haired Californian who joined the Royal Canadian Air Force in 1941

and transferred to the USAAF after Pearl Harbor. He graduated from flight school in February 1943 and was assigned to the 433rd Fighter Squadron in July.

Wire would be flying the tail-end position on the wing of Captain Herbert Jordan. When the word finally came down the pilots were told everything they needed to know; after takeoff the 433rd would circle the Lae area until they rendezvoused with the B-25s at ten minutes to ten and then they would head toward the familiar target of Wewak. Other squadrons would escort B-26s and B-17s to the Cape Gloucester area of New Britain.

The run to Wewak was routine with Wewak harbor itself half hidden under a curtain of broken clouds. Wire watched the olive drab shapes of the B-25s from his altitude of about 10,000 feet while they went in far below to attack the strip. He could not see any Japanese aircraft but heard over the radio from the bombers that they were coming in. The escort leader called out to drop external tanks and Wire felt the slight thud

432nd Engineering Section: Left; MSgt John Kobaly, his assistant, Cpl Frank Dea is in the center and the man on the far left is identified as Lt Krall.

475th Armament Section of one of the squadrons at Dobodura.

Maintenance on one of the turbos of a 433rd P-38. (Krane collection)

433rd crewchief in the cockpit of Jack Fisk's #191.

of his own tumbling away before he followed Jordan down into a spiral toward the bay.

At about four thousand feet Wire saw an enemy single-engine fighter coming in on Jordan's left. He turned just long enough to get a shot at the Oscar before it broke away from its attack on his leader. Then four other Japanese came down and zoomed up quickly behind Wire and Jordan who was now several hundred yards ahead of Wire.

Frantically, Wire tried to catch up with Jordan and nervously checked behind his own P-38; one dark silhouette of a Japanese fighter was closing in on his tail. All his training told Wire that the only chance he had was to form a team with Jordan that would drive the enemy off. He called to Jordan while breaking to the left himself, but apparently his leader could not hear him and Wire noticed to his dismay that there were two more enemy fighters above him.

The two Oscars rolled over on their backs and attempted to trap Wire in a split-s dive. Before they could complete their roll, the harried P-38 pilot did an extremely desperate thing and turned into them. Wire knew he was not supposed to maneuver with the more agile Japanese fighters, but he somehow got an accurate burst into one of the Oscars that shuddered from the sledgehammer blows of the P-38's guns and fell off into a spiral toward the water, trailing heavy smoke. The other Oscar prudently broke off contact and flew away.

All this time Jordan kept his wingman in view and gasped at the original Oscar blazing away at Wire's tail. The Japanese pilot must have thought that he shot Wire down or was shaken badly by the approach of Jordan's P-38, because he failed to culminate an easy victory. Wire finally lost his pursuer in a fast dive. Two more Oscars came at Wire after he ditched his original tormentor, one of them so close that Wire simply had to hold down the trigger and saw flame and smoke come from the enemy engine before the Oscar disappeared behind him.

At this time Wire was down to about 4,000 feet and joined up on Jordan's wing while the two of them were heading toward a transport in the harbor. What seemed to be a string of firecrackers going off above the ship turned out to be antiaircraft fire and Wire turned away to avoid it. The bombers had hit their target and were circling for home and, since his own ammunition was about gone, Wire decided it was time to head home.

Perhaps because of gun camera evidence Wire was credited with a definite Oscar shot down, another probably shot down and the third damaged. Ralph Cleage managed to get another probable and yet another Oscar (identified at the time as an A6M-3 Zero 32 "Hamp") was confirmed and a Dinah probably destroyed to give the 433rd an opening balance of two kills confirmed, three probables and a damaged for one P-38 missing.

John Smith's Journey

Second Lieutenant John Smith was the pilot of the missing P-38 as well as the victor of the Hamp and probable victor of the Dinah. He was considered one of the wildest pilots of the 433rd who developed a reputation for eagerness when he could be spotted rushing into a formation of Japanese aircraft regardless of odds. His bravery was impressive, but his squadronmates thought that he would surely be one of the early casualties.

He almost did become a casualty on the mission of September 2. Flying tail-end Charlie in Possum Blue Flight (Possum was the 433rd call sign) led by Lieutenant William Horton, Smith was at 5,000 feet watching the bombers go in on their run when about twenty five Oscars and Tonys came in from the direction of the sea. The Japanese started a dive on the B-25s and Horton led Possum Blue Flight down to the rescue.

Maintenance on 431st Squadron P-38 #126.

Smith vigorously accepted combat and soon made a head on pass at an Oscar that was attacking one of his flight's P-38s. It was the wild sort of confused combat that Smith liked best and he was all over the sky, joining up with Captain Joe McKeon at one point and shooting down an Oscar that had attacked McKeon.

The Oscar had pulled up over McKeon and turned to the right, giving Smith a good sixty degree deflection shot. Smith fired all the way to minimum range and watched the Japanese fighter explode into flames behind the pilot's seat. The blaze continued until the entire rear half of the fuselage was engulfed in fire and the aircraft fell out of control.

In the fighting that lasted for a good twenty minutes Smith hadn't realized that he had drifted south of Wewak. His ammunition was just about spent in heavily damaging a subsequent Dinah, so he decided to make for home. His compass was out and he had no maps of the area below, but Bena Bena radio assured him of the proper direction.

While he was making his way over the Lake Katabu area Smith managed to contact Captain Jordan and told him he thought he could make it to Marilinan. He then tightened up control of his P-38 and let down to 3,000 feet under the cloud cover.

432nd Squadron taxiing for takeoff. #165 was P-38 of "Rat" Ratajski who got his first victory over the Huon Gulf on the squadron's big day, September 22, 1943. (Krane collection)

From time to time the coastline was visible through the overcast and he got the unhappy feeling that he was flying the wrong heading toward the Finshhafen area. He made a ninety degree turn to the right in the hope that would bring him in line with a friendly strip, but after fifteen minutes it was obvious that he was hopelessly lost and heading nowhere useful.

Around noon there was very little gas left in his tanks and he called Bena Bena with a description of the terrain below and to inform them that he intended to crashland. He selected the most suitable field below, crossing it several times in preparation for the final wheels up landing.

Tightening his safety straps and cutting his switches he put his arm up against the armor glass for extra protection before he waited to touch down. For a few terrible seconds the wind screeched against the canopy, and then the P-38 crunched heavily into the ground and skidded like a dull plow until a branch smashing through the windshield announced the airplane coming to rest.

Smith was forever grateful that he decided to protect his head by placing his arm against the armor glass. Bits of tree limb and plexiglass had slashed his arm through the sleeve; better a bloody arm than a bloody face.

The crash must have dazed or even knocked him unconscious for a few seconds because the scene was quite still when he finally climbed out of the cockpit. He looked the P-38 over and decided that all he needed to do was change the radio settings and fire a few pistol rounds into the IFF (Identification, Friend or Foe) set. He also tried to retrieve the gun camera in the nose without success; even in the formidable circumstances that he found himself Smith was the eternal fighter pilot trying to insure a kill confirmation.

Since he had no idea where he was or even if he were in friendly or enemy territory he set off in a southeasterly direction. With all his necessary gear stored in a pouch made up of parachute strips, he walked for two days until he reached what he later realized was the Wassi Kussa River, off the Gulf of Papua and well to the south of any course he may have intended.

He decided to build a small raft to sail down the river to any settlement he should happen to find. Securing his parachute pouch to the raft he went to sleep for the night only to find that the tide had taken his raft and possessions when he awoke the next morning.

Now, with only a jungle knife and a small hand knife, he faced the disheartening prospect of following the river bank on foot. For seven days he made his way along the river until he finally reached the Gulf of Papua. The cuts on his arm and feet were swelling and making any progress extremely painful and difficult, and he had yet to find another human being on his trek.

On the next day – the ninth after his crash – he was discovered by an old native who took him to an island where food and medical treatment were available. He had not eaten for seven days and his hands and feet were badly swollen.

Above and left: Joe McKeon scored two victories for the 433rd on September 22, 1943 to bring his tally up to four. REGINA COELI means "Queen of the Skies." (McKeon)

Even though the treatment was primitive, Smith looked on it as merciful luxury.

For the next few days he recovered nicely and was able to walk and use his arms within limitations. The natives had told him that two Australian radio operators were on nearby Siabai and he persuaded them to take him there. Siabai was across the gulf from Port Moresby and seemed to offer an easy rescue by flying boat.

It took thirty-six hours to make the few miles by boat to Siabai, The natives made Smith comfortable on a bed in the bottom of the boat and even kept a fire going so he could be warmer and have a little tea. The Australian wireless operators were most helpful to the downed American and radioed for a PBY Catalina that picked him up on September 15 and took him back to Port Moresby.

Smith was admitted to the 171st Evacuation Hospital in Port Moresby and stayed there until September 21 when he was released to fly back to Dobodura. He reported to Major Low that same afternoon and wrote up his reports the next day.

One other 433rd pilot had a similar though less arduous experience than Smith. 1st Lieutenant Stanley Northrup had run out of gas while he was being pursued by a number of Japanese fighters during the Wewak raid of August 24. He crashlanded in the bush, but was able to merely walk back to Dobobura because this time the pilot was on a straighter course.

Huon Gulf

During Smith's ordeal the 475th had fought some of its fiercest battles to date, and was in the midst of a campaign that brought on some spirited opposition from the Japanese. Lae, Salamaua and Finschhafen with bases on New Britain had dominated the Huon Gulf since the Japanese landed there early in 1942, but the tide was turning in the favor of the Allies.

On September 5, 1943 the Australian 9th Division began to press onward to Lae and American paratroopers made their first appearance in New Guinea by landing on the nearby abandoned base of Nadzab. Pressure had been mounting on Salamaua to the south and by the eleventh of September it appeared that the Japanese would have to desert both bases.

If the Allies were successful in this operation only the bases on New Britain would be in position to challenge a complete sweep of the Huon peninsula south of Saidor. Even those New Britain bases would be largely powerless to enforce Japanese will in eastern New Guinea.

On the evening of September 3 there was a group meeting to lay out assignments for the covering of the Lae invasion convoy. The 433rd would cover destroyers and landing barges in the afternoon. John Babel attended the meeting and seemed impressed by the numbers; he counted about 100,000 men involved in the operation.

After an uneventful morning patrol the 433rd was off on an afternoon scramble at 1:30 on September 4. Bill Jeakle had a bit of bad luck when his canopy blew off right after takeoff and he had to land and watch the others fly off into the distance.

It didn't take long to reach station over the convoy at about two o'clock. It took even less time to sight the incoming enemy who showed up a bare twenty minutes later. There were an estimated thirty bright green Betty bombers coming in at around 12,000 feet and about fifteen single-engine fighters were with them. Above were eighteen more Japanese single-engine fighters at 20,000 feet.

The 433rd formation divided to make the most effective attack on the bombers. Most of the P-38s waded into the bombers at 12,000 feet with Herbert Jordan and Ken Kirshner downing two of them while several others were damaged and the bombing disrupted.

Joe McKeon was leading Blue Flight at 14,000 feet well in the trail of the squadron over Natter Bay, below Lae, when he saw Red and White Flights already engaging the bombers. When he approached the battle area he could see a number of Japanese fighters milling around below and led the flight to the attack.

With his wingman, 2nd Lieutenant Walter Reinhardt, following behind, McKeon selected one Oscar for a head on pass. The dark greenish-brown Oscar turned into McKeon's pass and immediately took devastating hits. Reinhardt was excited in his first big battle, but noticed the Oscar when it passed them burning and twisting in a steep dive toward the water.

Both McKeon and Reinhardt climbed and dived into the swirling fight and both fired brief bursts at fleeting targets before they were forced to break away. For about thirty minutes they engaged the Japanese until the fighting dissipated. Reinhardt saw a burning barge below indicating that some of the Japanese had hit their mark. Actually, at least three barges were hit and burning and a destroyer was damaged by a near miss.

The 433rd had done well in its second full engagement; in addition to the two bombers and McKeon's Oscar, 1st Lieutenant John Parkansky claimed another fighter and Lieutenant Dale Meyer probably destroyed yet one more. Four P-38s were damaged. Colonel Prentice led a 432nd Squadron flight earlier in the day and claimed a Zero, but Lt. Ryrholm was shot down and killed.

The next day the 433rd escorted a bombing raid to Cape Gloucester in the morning and flew a patrol covering ships and barges in the Morobe Bay-Hopoi area later in the afternoon. After an eventful few days for the squadron John Babel and Herb Jordan got together after dinner to celebrate Jordan's victory over the Japanese Betty bomber with a bottle of wine. It would be a rare moment for savoring a victory in the days to come.

On September 7 the 433rd escorted B-25s that bombed and strafed along the Markham Valley. During the mission there were calls from Morobe Harbor that the Japanese were bombing, but the 433rd was committed to the B-25 escort.

The 432nd Squadron was able to respond to the calls for help with four flights of P-38s. Captain Byars led the first flight and Noel Lundy led the second, Danny Roberts led the third and John Loisel had the rear flight.

Lieutenant Lundy was a bit irked at "Nephew", as the local fighter control was codenamed, for allowing the P-38 formation to lose altitude and fly in the wrong direction after the B-25s in the Markham Valley before the correct vector toward Arawe, New Britain was given. Nevertheless, Lundy's Green Flight was the first to sight the retreating Japanese and led the way to intercept them about sixty miles south of Arawe. Danny Roberts got an Oscar, Chris Bartlett also got an Oscar for his first kill, Lt. Tom Simms and Campbell Wilson each got an Oscar and Lundy himself claimed a Zero right over the harbor.

Return to Wewak

Even though Wewak had ceased to be an offensive threat to Allied forces moving northward, the base still received occasional visits by American bombing raids. The potential threat presented by a resurgent Wewak haunted Allied commanders and the few operational aircraft left at the base fanned that specter with an occasional surprise raid.

The 78th Sentai, for example, was nearing exhaustion and could only muster a handful of Tonys for any mission. Yet, Captain Nakahama was able to lead six of his 78th Tonys to counter the Allied landings near Lae. On September 13 he even led a harrying strike on Tsili Tsili.

That same day the Fifth Air Force sent a strike of B-24 bombers against Wewak. The 431st and 432nd Squadrons covered the bombers and ran into a whirlwind of a fight involving the few Tonys left at Wewak. By the time it was through the fight had claimed about six Oscars, two Tonys and a Nick, but at a cost of three P-38s shot down and two pilots killed.

The 431st Squadron sent off fourteen P-38s at eight o'clock in the morning. Three planes aborted before the rendezvous with the B-24s at 9:40, but the remainder proceeded on to the target and sighted the enemy just south of Boram strip.

About fifteen Oscars and a single Tony were identified at an altitude of 12,000 feet. The P-38s were at 18,000 feet and in good position to attack the enemy from the most advantageous quarter. After maneuvering for about five minutes the 431st came down like heavenly wrath and the Japanese formations weaved like mad to counter the American attack.

Captain Verl Jett went down into the melee with Frank Lent on his wing. The Tony made a head on pass at the formation, but Jett and Lent lost it in the confusion of the battle. Most of the 431st flights became scattered when everybody wanted to get on the tails of the wildly turning Oscars.

Jett saw a number of Tonys below and led Lent down to the attack. Two flights of P-38s were already making a pass at the Tonys and Jett noticed a twin-engine Nick on the tail of the trailing P-38s. It was a set up for Jett even though there were smoke trails coming from what must have been a rear gun on the Nick. A few bursts from the P-38 and the Nick exploded into flame just as Jett passed over. Lent saw a parachute from what must have been the gunner's seat blossom over the falling enemy plane.

Tom McGuire became frustrated when he couldn't drop his tanks to join the battle. He radioed his wingman, Lieutenant Vincent Elliott, to join up with another flight. Elliott would have been glad to oblige, but every other P-38 had since disappeared into the countryside below, chasing the enemy!

But McGuire was already heading inland and Elliott was eager enough not to retire from a battle while he still had a perfectly good P-38. He began searching the skies and horizon for any sign of someone to join or someone to attack.

Within a short time he found three P-38s south of Boram strip diving on four single-engine fighters. He joined the chase and got two good bursts from about sixty degrees deflection on an unlucky Oscar. Pieces of the Japanese plane flew off into the slipstream and a trail of light-colored smoke followed the Oscar down into a spin. Elliott watched the plane fall below 3,000 feet before he decided it was not prudent to go any lower.

Blue Flight, consisting of Captain Haning and Lieutenants John Knox and Don Bellows, was separated from the rest of the squadron and was heading back when a plot was called in that took it south of Madang and north of the Ramu River. Four Tonys and six single-engine fighters were simply circling the area.

Groundcrew of 433rd P-38 #192.

Cleaning the barrels of what appears to be Frank Lent's first P-38. Before the placement and arrangement of squadron numbers was made standard on the P-38 nose, the 431st numbers were often placed beneath the guns ala 39th Sq. practice

Haning lost no time in attacking one Tony that proved how agile it was by quickly getting on the P-38's tail. Bellows and Knox were right there to rescue Haning, but two of the single-engine fighters and another Tony deftly got on their tails.

Captain Norbert Ruff of the 80th Fighter Squadron witnessed the combat and confirmed the crash of the Japanese fighter.

In the tight turns that followed, Don Bellows blacked out for a few seconds. When he came to his senses he was shocked to find his P-38 badly shot up and on fire. He called out his position over the radio and marvelled that no other fighters – Japanese or his own flight – were in sight. He crashlanded about a mile east of the Ramu River.

Haning had escaped the Japanese, somehow, and landed at Tsili Tsili. He met Captain John Hood who was refueling his own P-38 there and told the story. Hood took off and searched for Bellows whom he found standing near his burned P-38. The next day a Piper L-1 landed near the crash site and rescued Bellows. Knox was never seen again.

While Hades Blue Flight was getting the worst of it from the Japanese, the 432nd was taking some damage, too. Major Tomkins led sixteen P-38s to the target area and got into the scrap a few minutes after the 431st. The bombers had just come off the target and were heading home when Tomkins noticed a pair of single-engine fighters just below him. He made a head on pass on one of them and pressed the attack to within twenty feet, noticing uncontrollable flames coming out of the engine cowling of his falling victim.

Noel Lundy was leading the second flight and followed Tomkins down to the attack. Within the next few minutes Lundy observed a pair of Tonys split-s from a cloud down on the tails of two P-38s. Lieutenant John Weber was Lundy's wingman and watched him fire at the Tony.

Far behind, Lundy's second element could see the entire combat, also. 2nd Lieutenant Paul Lucas was bringing up the rear and saw "cannon shells bursting on the wings and fuselage of the Tony from Lt. Lundy's fire. The Tony did a half-roll to the right and went into a sharp, flat spin." Lucas's element subsequently lost Lundy and Weber and joined up with Major Tomkins to fly down to Marilinan for gas.

Lundy and Weber were returning home when three Tonys jumped them near the base at Dumpu, about thirty miles south of Astrolabe Bay. Weber managed to escape the attack, but last saw two of the Tonys raking the belly of Lundy's P-38 with accurate fire.*

One subsequent attack by Tonys on the fourth 432nd Squadron flight near the headwaters of the Markham River at 18,000 feet was beaten off. Lieutenant Bill Ivey of group headquarters, who was leading the second element, shot down one of the Tonys.

It was difficult for the 68th and 78th Sentais to muster any more than about twelve Tonys between them during this period of the New Guinea struggle. By September 19, 1943 it was necessary to relieve the 78th Sentai briefly until replacement aircraft could be delivered from the Philippines. If the two Tony units could send up no more than ten to twelve interceptors on the September 13 engagement then it is a remarkable tribute to their flexibility that so many 475th formations could have identified them over so wide a battlefield.

Moreover, it was becoming more apparent that the Kawasaki Ki-61 Tony was the most dangerous Japanese fighter in New Guinea. Most of the P-38s verified lost in aerial combat seem to have been victims of Tony fighters. P-38 pilots who met the Tony in battle were generally of the opinion that the Japanese fighter was a bit faster than other types, but was still easy to catch and defeat.

September Lull

The next few days were filled with dull patrols and escorts. The Japanese were forced to regroup their aircraft while the invasion in the Huon Gulf gradually wrested one base after another from them. John Babel's diary of the period makes this laconic note of September 17: "Flew patrol at Morobe Harbor while destroyers & barges moved along the coast. Allies have captured Lae and 14,000 Jap troops are retreating through the mountains toward Madang."

_____
*NOTE: Lundy's body was found in January 1944 and he was buried near Bena Bena.

Verl Jett claimed his fifth Japanese victory on September 13, 1943 (475th FG)

So it was for the next couple of days; patrols over Morobe, weather reconnaissance or even routine intercept scrambles that found only phantom plots. In truth, the pilots were glad for the break in action. Many of the original group pilots had agreed to stay on for an extra six months beyond their expected rotation date, and every single day brought them closer to a permanent trip home.

There was one break in the respite that proved again the dangerous potential of the Tony fighter. On September 20 there was another escort of B-24s to Wewak that was uneventful until the 432nd Squadron escort began seeing smoke or dust on the ground near Nadzab and heard the uncomfortable call from fighter controller "Shadow" of bandits in the area. Shadow was almost directly below and in visual contact with the ensuing battle.

Captain Jim Ince almost immediately made a turn in the opposite direction with his wingman, Lieutenant Billy Gresham. Only the second element of the fourth flight, Lts. Wilson and Peregoy, had turned, also, and they were being attacked from out of the clouds by as many as six Tonys!

The Japanese had achieved complete surprise and hit the 432nd formation with overwhelming suddenness. Jim Ince led Gresham to rescue of Wilson and Peregoy, but a burned out connection to his guns forced him to retire and head for home. Gresham carried on alone.

Peregoy's P-38 was hit in the engine and he was forced out of the fight. Gresham was climbing to the aid of his comrades and perhaps distracted the Japanese long enough for Peregoy to escape. However, three Tonys were able to get above and behind Gresham and forced him to roll over and dive away.

At some point in his dive Gresham started losing rivets from around the left intercooler; his indicated airspeed was pressing 600 miles per hour. Fortunately, he lost his pursuers and came up behind another pair of Tonys that probably never saw him. He fired a few bursts after waiting until he was within about one hundred yards of the lead Tony and hit it so hard that oil flew back from the doomed airplane and slapped right on the P-38's windshield.

Somehow, the other Tony managed to get on Gresham's tail, but was unable to fire, perhaps because it was out of ammunition. It left Gresham's P-38 while the first Tony was observed by ground personnel as well as Gresham to crash about three miles north of Shadow.

Gresham was an aggressive fighter pilot and not one to simply let an enemy plane skulk off. Once again the P-38 drew up into close range of its victim and poured a devastating line of gunfire at the Tony. Bullets exploded the length of the fuselage and black smoke trailed back from the Tony's underside radiator. Detonating cannon shells tore large pieces off the tail just before Gresham overflew the wreckage.

Other Tonys were maneuvering for position on Gresham's tail. He was low on ammunition and both intercoolers on his P-38 were out of commission, so he dove out of the fight and headed toward Shadow where he lost his last pursuer. In the distance Gresham could see an Oscar falling in flames with both wings coming off in the spin. He learned later that LeRoy Bryan of the first element had shot it down.

The 432nd was lucky to have turned the situation to their advantage. The final awarded claims were two Tonys and one "Hap", as it was identified, destroyed and another Tony probably destroyed. The Tony pilots were rated as aggressive but inexperienced; they were almost certainly from the replenished 68th Sentai, which was now operating some Oscars to supplement its Tony force.

Besides Peregoy's damaged P-38 432nd accounts list Lieutenant Tom Simms P-38 as being lost at the same time. He was observed bailing out, but was never recovered.

Record-breaking Over Finschhafen

The Japanese Navy had been ignoring New Guinea for what it considered the greater threat in the Solomons. By the middle of 1943 Guadalcanal was in Allied hands and the arduous New Georgia campaign that was begun in July was culminated in August. Vella Lavella was assaulted in the same month that

New Georgia fell and 10,000 Japanese troops were trapped on the central Solomons Island of Kolombangara.

What the Japanese Army and Navy both hoped to accomplish was complete protection of all Imperial conquests. It was a catastrophic shock that the Americans and Australians could not only manage to check Japanese advances but were rolling back over one base after another with what seemed to be alarming ease.

The fall of Lae and the imminent threat to Finschhafen by the last weeks of September caused something near panic in Japanese naval headquarters in Rabaul. The Allied success in the Solomons threatened the Japanese linchpin at Truk in the Central Pacific, but the loss of Finschhafen meant the possible isolation of Rabaul with a resultant steamroller over northern New Guinea that would not end short of the vital oil conquests of the Indies.

Improving weather over northeastern New Guinea in the latter half of September prompted the Japanese 11th Air Fleet on New Britain to schedule air attacks on shipping traffic heading up the Huon Gulf. The 751st Kokutai sent off four Bettys from Vunakanau Field, Rabaul, on the night of September 21 while five more Bettys of the 702nd Kokutai took off during the early morning hours of September 22.

While neither force managed to find the Finschhafen invasion force and were obliged to bomb secondary targets,the 751st Bettys sighted a transport convoy near Cape Cretin. When the Japanese air command was informed it hastily assembled a strike force of eight 751st Bettys armed with torpedoes and twenty-three Zeros of the 253rd Kokutai with about fifteen others from the 201st and 204th Kokutais.

The Bettys took off in three waves; the first of three bombers led by Lt.jg. Yada, the second of three led by NCO Jitsuyoshi Kuramasu and the last two by Chief Flight Petty Officer Aoki. Takeoff time was 8:40 in the morning amidst a flurry of cheers and waving hats from the ground crews. The bombers circled the field once and then headed down the length of New Britain toward the target. Twenty Zeros take up position close above and behind the bombers, but the bomber crews know that the chances of surviving a daylight torpedo mission are not very good.

An Australian wing commander suggested a solution to the problem of advanced fighter control during the Huon Gulf landings. Capable officers would be placed aboard a destroyer in the proximity of the invasion area and call out plots as they progressed toward the convoys. Covering fighters would be tuned into the destroyer controller frequency for the latest sightings.

This fighter direction procedure worked well during the first landings near Lae when the destroyers Reid and

Conyngham were used. The Reid again acted as controller during the Finschhafen landing, this time remaining close to the convoy to coordinate the large number of fighters that were scheduled in the cover force.

The landings began at 4:45 in the morning at SCARLET beach near the Song River just north of Finschhafen. Australians landing on the beach made swift progress and quickly captured the key airstrip. American air attacks continued with what were considered to be good results against installations in Finschhafen itself despite cloudy conditions.

With an apparently completely successful operation at hand the convoy pulled up anchors and headed back toward Buna shortly before noon. There had been very little air opposition up until now, the only action reported by the 433rd Squadron that happened to be patrolling between Lae and Finschafen. Joe McKeon heard the controller call out the bogie and led his Yellow Flight from high altitude to catch a lone Tony at 27,000 feet. McKeon made sure of the target then had to fly through the ball of fire that was left after he exploded the Tony.

Before the noon hour was over, however, the Reid began getting large numbers of plots coming from the direction of New Britain, less than seventy miles from the convoy. Three American fighter squadrons that were covering the convoy, the 341st with P-47s, the 35th with P-40s and the 39th with P-38s, were due for relief within a few minutes, but all had sufficient fuel to adequately engage the oncoming Japanese force.

Two relieving squadrons, the 9th and 432nd, were already in the air and could be vectored toward the Japanese force. Fred Harris was leading the 432nd at about 12,000 feet, but could not find the enemy force and DUCKBUTT, as the controller on the Reid was codenamed, was unable to give an altitude. Five American fighter squadrons were circling like hungry vultures, each pilot eager to get at the promised feast below.

Then Duckbutt called out, "bandits in sight from 6,000 feet on down." Harris identified six Bettys covered by about ten Zeros below. The haze and reflection off the water must have distorted the visibility somewhat because the bombers looked reddish-brown to him and the Zeros seemed to be camouflaged an olive drab. One of the Bettys also seemed to have a pair of reddish diagonal stripes on the fuselage.

There were also between twenty to thirty Zeros well above the 432nd P-38s. The Japanese were rolling and looping to keep the Americans in sight, but Harris decided to ignore them and ordered drop tanks released before heading down to cut off the bombers before they reached the convoy. Harris led the squadron down in a fast dive that was thwarted when the Bettys managed to slip beneath a thin layer of clouds.

The 432nd Lightnings pulled up over the cloud deck and ran smack into the Zero close escort. At the same time the Japanese high cover had arrived from above and behind. Harris looked around and saw one of the Zeros slip in uncomfortably close on his tail. Harris's wingman, Zach Dean, saw the Zero begin to get some hits on his leader's P-38, but easily got in behind and shot up the Japanese fighter which disintegrated and fell into the sea.

Harris could see that the Zero cover attacking his flight had left the bombers open to attacks by the last two flights of P-38s led by Jim Ince and John Loisel. He managed to clear his flight long enough to get on the tails of the bombers himself and led Dean up behind a pair of them. The one Harris attacked was observed to catch fire and fall into the sea. Harris saw the Betty that Dean fired at explode in mid-air.

Lieutenant Vivian Cloud was leading the second element in Red Flight with Don Garrison close on his wing. The third Betty in the lead wave was already flying into the black puffs of antiaircraft fire being sent up by the convoy. Cloud fired one burst that knocked out the tail gun, then another that silenced the dorsal position. Without return fire from the bomber the P-38 was able to draw up directly behind and blaze away until the pattern of strikes ripped up the fuselage and started the left engine burning.

When Cloud overflew the falling bomber he immediately saw one of the escorting Zeros directly ahead. Surprisingly, the Japanese fighter seemed to be making a slow roll for no apparent reason. Cloud lost no time in taking advantage of the moment and closed in to fire when the Zero reached the top of its roll. Bullets slammed into the left side of the cockpit and ripped through to the right wingtip before the Zero fell off and dropped straight into the water.

Below him, Cloud could see four enemy aircraft burning on the water. There was also a P-38 out in front getting rough treatment from two Zeros who were getting hits that set an engine on fire. Cloud couldn't see the number on the P-38 in distress, but it was probably his own wingman, Don Garrison, who was flying #151 that day.

Four more Zeros came down on Cloud's tail while he tried to help his fellow P-38 pilot. The four Zeros dived down from the left and Cloud went into a dive that seemed to lose his attackers.

However, the P-38 did not escape serious damage when its right engine burst into flames that Cloud could not stop. The fire was spreading so Cloud released his canopy and seatbelt and struggled out of the cockpit.

Hanging in his parachute straps within the combat area was not where Cloud wanted to be so he released himself and fell into the water as soon as possible. Splashing around in the Huon Gulf made him realize that his rubber raft was still attached to the parachute, obliging him to retrieve it before he could have the relative safety of the little dinghy. Within twenty minutes a destroyer picked him up, but he later learned that Garrison was indeed missing.

Captain Waldman's Clover White Flight had run directly into the Zero top cover that was attacking the lead flight. After the first head on pass White Flight was struggling to reform and Waldman's wingman, 2nd Lieutenant Howard Hedrick, fell back behind Flight Officer Charles Ratajski in the White four position when Waldman dove away.

While Hedrick tried to catch up to his flight he could see three more Bettys below (probably the second wave led by Kuramasu) and that Waldman was well out ahead and alone. The Zero popped up long enough for Hedrick to get a brief shot at it before joining Waldman. Another Zero was obliging enough to let Hedrick get into close range (100 yards) and fire a five-second burst. The Zero turned and let Hedrick get in another deflection shot that sent the enemy fighter cartwheeling into the water.

Hedrick and Waldman then had bad luck in attacking the bombers when Hedrick's gunsight went out and Waldman's guns jammed. Ratajski had similar fortune when he attacked two Bettys in succession without results. His element leader, Lieutenant John Rundell did a little better and claimed one of the Bettys shot down.

Ratajski had watched what he believed to be bombs being jettisoned from the Bettys before they turned for home and his flight attacked them. After his last pass at the bombers he lost the flight and headed toward the land. He climbed to about 3,000 feet and saw a P-40 stalking a Zero. Since he had the altitude, he dived down and beat the P-40 as well as Clover Green Flight to the Zero that was flying low and slow over the water. A single burst of fire into the canopy was all that was needed to send the Zero down splashing.

At least two of the Bettys were seen by Kuramasu to fall in flames into the sea. His own bomber is quickly attacked by a P-38 and a P-40, making it impossible for him to observe the results of his own torpedo run. There are no Zeros visible to give aid to the single retreating Japanese bomber, so the only hope is to dive from ragged cloud patch to ragged cloud patch, keeping the American fighters away from a tail shot with the skill of the 20mm cannon gunner in the rear.

The Betty is shot full of holes, but the resolute tail gunner continues to defend the rear quarter in spite of some serious wounds. It would seem that this lone Betty was claimed as destroyed by more than one American as the damaged bomber disappeared under one cloudbank after another.

This Betty was the only one to survive the battle when it made a harrowing landing at Cape Gloucester. Ironically, even the one surviving bomber of the mission was destroyed on the ground at Rabaul during the American bombing raid of October 12, 1943.

About eight Zeros were reported shot down, mostly from the 201st Kokutai. The American units were exultant about the results which netted them a total of forty claims. The P-38s claimed twenty-nine for the loss of two 432nd Lightings, while the P-47s claimed four and the P-40 claimed seven.

John Loisel had already led Clover Green Flight through some heavy action against the Zero escort and had exploded one that simply tumbled all the way to the water. LeRoy Bryan was Loisel's wingman and claimed two Zeros himself before he and his leader had all but exhausted their ammunition. When Ratajski got his Zero just off the water near the shore, only Clover Green Three, Flight Officer Henry Condon had any appreciable ammunition left. As it happened, another Zero was in the same area and Condon shot it down in a head on pass.

All this time the leader of the second wave of Bettys, Jitsuyoshi Karamasu, had been pressing home the attack and witnessed much of the combat. He saw the ships of the convoy – about eleven transports and destroyers in all – breaking formation and making large white wakes in the jewel-blue sea. All around the close Zero escort was zooming and speeding ahead to meet the P-40s and P-38s now visible to Karamasu's crew.

Karamasu himself concentrated on the largest ship he could find and was only briefly distracted by the large water-spouts and black puffs of bursting antiaircraft fire. The observer in the nose of the Betty cries out the distance and finally orders the release of the torpedo. The Betty heaves and skids to avoid the lacing machine gun fire while Kuramasu pulls up over the stern of his target.

Even though some of these claims are certainly removed from the specific action against the convoy it is easy to see that there was a good deal of overclaiming. Nine Bettys were awarded to individual pilots as confirmed when no more than eight were involved in the mission and one of those made it home, albeit in a heavily damaged condition.

Eighteen Zeros were identified as confirmed victories, while Japanese losses were limited to eight. Even discounting the stray Oscar that could have wandered close enough to the battle area to have been mistaken for a Zero victory there must have been a number of examples of pilots making duplicate claims.

Sgt George Heap was already a distinguished hero before he came into the 432nd FS. In 1942 with the 8th Fighter Group, he had led a reconnaissance team under the very noses of the Japanese and directed by radio the destruction of an enemy flotilla heading for Milne Bay. (Heap via Maxwell)

Whatever the truth of the confusion that attended this mammoth engagement the 475th enjoyed a new record for the most victories confirmed for a single squadron in a single engagement. The 432nd was finally awarded seven Bettys and eleven Zeros shot down. Squareloop Harris also claimed two Zeros in addition to the Betty to become the latest ace of the group.

Chapter V

ORO BAY

Captain Dennis Glen Cooper had taken over the intelligence officer duties for the 431st Fighter Squadron by the first weeks of September 1943. He was a bit older than most of the officers of the squadron, being in his late thirties by the time he joined the 475th. He had used the time industriously, nonetheless, getting a Bachelor of Arts in Geography from Wayne University and winning a Phi Beta Kappa key in the process. He was accepted into various scientific societies and wrote a number of articles for their journals.

By the time he had applied for a commission in Army Air Intelligence he was head of the science department of a Detroit high school and was working on his Ph.D. Cooper became known as a dynamo in 475th air intelligence and his alert hut became the most elaborate in New Guinea, by reputation.

It was probably Cooper's devotion to his pilots and crews that really made him extraordinary. Years after the war he referred to them as "my boys", with almost fatherly interest. He grieved when one of them was lost and felt genuine pain when he learned of troubles after the war.

Some of the 431st veterans fell on hard times when peace came with health or marital problems, or even that wretched bane of retired fighter pilots; alcohol. Cooper kept in touch with as many men as possible through the 475th Association.

But his value to the men of the 431st during the war was manifold. As soon as a mission was in the air and security was lifted, Cooper would pass all information along to the groundcrews and update them on the events, especially any contact with the enemy. During a few days in October 1943 it would not be necessary

Dobodura Notes

Tsgt Mel Allan had a time of it in the early hour pre-flight checks. When he tried out the superchargers he had to stand his whole weight on the brake pedals and brace his back against the seat to keep the P-38H from moving forward. The side windows had to be kept open to detect every irregular sound and the scream of the turbine blades was deafening. The slip of air from the propellers didn't help much and earplugs weren't used at that time, all of which made the necessary checks an agony for the groundcrewman.

Number 116 was Allan's regular P-38 and several alternate pilots like Donno Bellows, Lowell Lutton, Paul Smith, and Alvin Kidd used it before Lieutenant John Tilley was assigned as the regular pilot. Allan had hoped to keep Lieutenant Kidd as a pilot long enough to capitalize

OPPOSITE: L. to R.: Donald G. Revenaugh, Donald Y. King and John Smith holding up fingers to denote the fact that they each scored victories over Oro Bay on October 15, 1943. (Hanks)

TSgt Bob Applewhite, the crewchief of McGuire's first two P-38s.

on his name for several planned pirate motifs, but Tilley was a good guy who went an extra step to accommodate his groundcrew.

Actually, it was good policy to keep the groundcrew in the best possible humor, since they literally had the safety of the pilot in their hands. Tilley would confer with Allan and the other men and help out with minor jobs necessary for the maintenance of his P-38. That is, he did these things until somebody up the line found out and instructed him to maintain distance from the enlisted men. Tilley and Allan shrugged it off and did what they could.

The 49th veterans like Allan took to wearing red-painted caps as a warning to newer men that they were old hands with about two years of service in the jungle. The identifying cap was supposed to let others know that the wearer had become mean in the New Guinea bush and was not to be trifled with.

475th officer's club at Dobodura. "Pappy" Cline is standing to left rear and Jack Mankin is sitting at bar with hat and jacket. (Brown)

Dobodura was probably slightly more comfortable than some of the Port Moresby bases because it was closer to the ocean breezes, but it was also more remote. The 431st crews had acquired some creature comforts like overstuffed leather chairs and other furniture in Port Moresby, but the logistics of transporting them proved insuperable and the disappointed men ended up at their new base with not much more than they could carry on their backs.

Consequently, life at Dobodura possessed a high level of boredom. One of the things that cut through the monotony was an item known as a jungle juice still that produced moonshine from various fruits and vegetables of the nearby bush. Some of the 431st crewmen, including TSgt Allan, constructed the illegal stills and kept them out of sight by camouflaging them near personal trenches.

One unbearably hot day a blinding downpour kept the 431st crews sweltering in their tents with only a prime batch of jungle juice for diversion. Mel Allan had stripped down to the buff both to dry off his work fatigues that got drenched before the flight line was closed down for the day and to escape the wretched heat.

Before he knew it, he was out in the blinding rain, riding a bicycle and whooping like a banshee. The constant sipping of the jungle juice had robbed him of his normal senses; by the time he woke up he was buck naked on the ground in the middle of the camp, wondering what had happened.

Penalties for the illegal stills were severe. One man got six months in the guardhouse after his third warning. Allan nearly went the same route when a surprise inspection of the trenches caught him flatfooted. He hurriedly covered the still with a stray piece of paper from the trench and waited for the inspecting lieutenant to pass by.

Allan stood by nervously while the serious looking young officer approached and leaned over to make an inspection of the renegade corner. Allan took his first good look at the piece of paper and noticed the dark spots for the first time. The officer gave him a sternly reproving look before he ordered the mess cleaned up and then moved on to the next trench.

433rd Transitions

Two days after the big fight of September 22 the 433rd was covering B-25s on a hunt for Japanese barges north of Finschhafen. Joe McKeon was leading Possum White Flight when he saw about nine Bettys covered by single-engine fighters at 10,000 feet. He led his flight in a climb behind Possum Red Flight from Possum White's own altitude of 3,000 feet.

#193 after Major MacDonald crashlanded on October 15, 1943. MacDonald disputes the official record that he met the Val bombers five minutes after takeoff and that he was assisted by numerous other American fighters including a P-40 that escorted him home. He was in the midst of the enemy almost before his landing gear was up and never saw another U.S. fighter until he crashlanded. (Hanks)

The P-38s managed to get on the tails of the Bettys, but the Japanese fighters were all over them before they could do much about it. McKeon saw his own wingman, Lieutenant Ray Corrigan, going down with both engines smoking. For the next few minutes there was a jumble of turning airplanes and McKeon sent one Japanese down in flames into the undercast, he also managed to shoot a large hole in the wing of a second target.

Ralph Cleage also managed to badly damage what he identified as a Hamp and was awarded a probable. A second P-38, piloted by Ken Kirshner, was missing after the battle. After the war it was discovered that he survived and was captured, but died in captivity at Rabaul.

Apparently the nine Betty bombers were able to reach their target and do some considerable damage to the Allied advanced headquarters at Finschhafen. Two days later the 433rd was more evenly matched during an escort of eleven B-24s to But and Dagua airdromes. John Babel, Clarence

Wilmarth and Martin Low all sent Tonys down to crash, but a heavy overcast at 3,000 feet denied them confirmed claims. Wilmarth was also credited with an Oscar destroyed and Cal Wire capped the day by confirming two Oscars.

The 431st escorted B-24s back to Wewak on September 28 and were surprised by a number of Tonys and Oscars at 24,000 feet. Captain John Hood shot one Oscar off the tail of Lieutenant Fred Champlin then shot down another attacking Tony. Champlin himself got his first kill when he downed an Oscar. Warren Lewis earned his second victory when he also shot down an Oscar.

Tom McGuire remained the hottest pilot in the 475th when he recorded on his gun camera film the destruction of two more Oscars. He was now the top-scoring pilot of the entire group with nine confirmed victories.

Two days later Henry Condon of the 432nd Squadron topped off the victories for September by downing another Betty off Finschhafen. In less than two months of combat the

475th had confirmed more than ninety aerial victories for ten combat losses.

By October 2 the Finschhafen area was declared secure. If the Japanese were staring blankly at their conquests being rolled back on them, then the Allies had mixed feelings about the greater ease with which they moved in central New Guinea and the formidable task of dislodging an increasingly desperate and consolidated enemy.

Many Japanese troops had escaped in the direction of Alexishafen and were determined to stand against any assault. The air action around the Huon Gulf, moreover, had prevented full interdiction of Japanese supply lines that were now more often being implemented by small barges and submarine transports. Upcoming Allied projects also included conquering the Bismarck Archipelago which presented a threat to MacArthur's drive on the northern coast of New Guinea.

The fall of Finschhafen did present a moment of celebra-tion for the 433rd Squadron. 40th Squadron veterans Herb Jordan, Bob Schick, Clarence Wilmarth and Major Low had received orders to go home after each had spent about eighteen months in New Guinea combat. None of them seemed sorry to leave for the U.S.

Wilmarth stopped by the flightline to say goodbye to Ted Hanks and to thank him for taking such excellent care of his airplane. Hanks would always think of Captain Clarence Wilmarth as a fine gentleman and great P-38 pilot until Wilmarth died of a heart attack in 1984.

Perhaps Wilmarth recommended Hanks to his friend and aviation cadet classmate, Danny Roberts, when Roberts took over the 433rd Squadron on October 3, 1943. Roberts selected Hanks to be his crew chief and the old airplane mechanic was immediately flushed with pride. It was a matter of high prestige to be selected to crew the "Old Man's" plane, and Captain Roberts was an ace to boot.

It was the beginning of a mutual admiration that lasts to the present in Hanks' veneration of Roberts's memory. Whatever deficiency crewmen like Hanks perceived in Major Low or other 433rd Headquarters personnel were more than elimi-

Virgil Hagan just 15 minutes before he was killed on October 17, 1943. He destroyed one Val and damaged another on October 15 and described the opening moments of the battle in a letter home on October 16: "We (had) just got off the field & saw their planes & bomb splashes. We dropped our belly tanks & started into them. For about a half hour one could see at least one Jap plane streaking out of the sky, & there were oil fires on the water." (via Jack Cook)

Fred Champlin in the cockpit of his trusty #124. (Cooper)

Champlin's #124 coming to grief on October 17, 1943 with Lt. Pare at the controls. Most 475th pilots were assigned their own P-38s as they became more available, but every mission assigned the most operational fighter to those who happened to be scheduled for the day. (Champlin and Ethell)

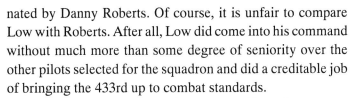

nated by Danny Roberts. Of course, it is unfair to compare Low with Roberts. After all, Low did come into his command without much more than some degree of seniority over the other pilots selected for the squadron and did a creditable job of bringing the 433rd up to combat standards.

Roberts, on the other hand, was something of a management genius who could get the best from his men by fully understanding them and leading them with the example of hard work. His diary for September 26, 1943 contains the following entry: "Today the Col. asked me to take a squadron and whip it into shape . . . This is a break but it will be hard work."

John Babel noted in his diary that on October 4, the 433rd flew a B-25 escort to some islands north of New Britain, then had its first meeting with its new commander during the evening. Roberts quickly impressed the pilots with his enthusiasm for their survival as well as maximum efficiency in accomplishing the mission. "Stay together like a pack of wolves", he would repeat, ". . . like a pack of wolves."

The first mission that Danny Roberts led was probably the initial Rabaul effort on October 12, 1943. Ted Hanks remembers that it was still dark on the flightline when nineteen P-38s

were readied and assembled for their pilots. The P-38 that Roberts would use, number 197, was at the far end of the line with Ted Hanks standing by. He could see Roberts at the far end of the line, speaking or nodding to everybody along the way. When he reached his own plane, Hanks stood up to attention and he greeted his crewchief with the warm smile that would become his trademark. By the time he took off his crewchief had become his most enthusiastic supporter.

Lessons Learned

By the end of September the already extensive combat experience of 475th pilots had been expanded by their operations together over the past two months. There were still some hot fighter pilots who tended to believe that the best way to get kills was go it alone and take the risk of shooting down an enemy plane before other Japanese pilots could arrive within attacking range.

Such pilots in the 475th were either dead or looking for another unit before the end of 1943. The new pilots coming in were drilled over and over again the way to survive and do

Above and right: Groundcrew posing near the 433rd commander's aircraft, probably late September or early October before Danny Roberts took over. (Krane collection)

some damage to the enemy was to follow the advice of the older pilots.

Basically, the advice on tactics broke down to six principles:

1. Stay in formation at least two planes.
2. Look around, keep your "head on a swivel."
3. Maintain high airspeed.
4. Think one move ahead know what to do after the next move.
5. Surprise on strafing raids with no more than one or two passes.
6. Make low, erratic and fast withdrawals.

Meryl Smith made up a questionnaire to check each pilot's knowledge of the P-38. Besides knowing what to do with the fighter in combat it was necessary to be an expert in flying the P-38. The more a pilot knew about fuel and hydraulic systems, emergency procedures and weapon systems the better chance he had of defeating a capable enemy pilot who was flying a highly operational Japanese fighter like the Zero.

The questionnaire would, for example, quiz the pilot about cannon stoppage in which the explosive round is struck with insufficient energy to fire. Ammunition in damaged or deteriorated condition was being received in the Southwest Pacific and P-38 pilots in the 475th Group were made aware of the situation in order to give them that much more edge in combat.

One of the reasons the P-38 was more effective in the Pacific than in same European units was precisely because of the type of care given to indoctrinate pilots that the 475th practiced.

October 15, 1943

If the Japanese had been nervous about Allied advances up the Huon peninsula, they must have been nothing less than edgy about the first escorted daylight raid on Rabaul. Pressure was mounting to strike back at growing Allied forces on New Guinea because the Japanese believed that the next invasion would occur somewhere in New Britain.

The first raid on Rabaul was enough to stun the Japanese. 115 B-25s, 100 B-24s, thirteen Beaufighters and 120 P-38s had taken off on October 12 and had done a good deal of damage to docks and warehouses and parked aircraft. Ralph Cleage had claimed an Oscar in the air, and John Fogarty, flying with the First Provisional Squadron claimed a Betty. Japanese Navy records list two fighters lost, nine medium bombers set afire on the ground and thirty-six medium bombers damaged.* Also, about one thousand drums of fuel were set afire and four ships were sunk.

During the night of October 13 and morning of October 14, photographic reconnaissance aircraft recorded what seemed to be a heavy influx of aircraft at Rabaul. Estimates ran as high

---

* NOTE: Translation of Japanese Navy chart showing air operations, Document No. 60881

as three-hundred bombers and fighters on the airdromes.

It was an ominous portent that was realized on October 15. The Japanese had decided to attack shipping at Oro Bay, which was a reasonable choice considering that supplies and troops, including a U.S. Marine division assigned to MacArthur, were staging through the base. A decisive blow could reasonably be expected to throw Allied plans into confusion, at least for the moment.

Fifteen dive bombers of the 582nd Kokutai led by Lieutenant Ikeda and thirty-nine Zeros of the 204th Kokutai under W/O Aoki left Rabaul for Oro Bay in the early morning of October 15, 1943. It was hoped that the crowded shipping in the bay would be taken by surprise and devastated before defenses could respond.

This was one mission that Cooper would not have to describe to the 431st groundcrews. Buna, the approximate destination of the Japanese raiders, was on a line from Dobodura about eight miles away. Anyone at the Dobodura camps would have an excellent view of the sky where the main battle would be fought.

The first scramble order came a few minutes after eight o'clock in the morning. Major Charles MacDonald was just driving up to Group Operations in his jeep when Captain Bill Ivey came running out and shouted out the news about the scramble. The two pilots then raced at breakneck speed to find a pair of P-38s. They tried the 432nd Squadron with no luck, but found some P-38s primed for takeoff on the 433rd line.

MacDonald got into Cal Wire's number 193 and headed for a quick takeoff. Wire was furious later on with his crewchief for letting someone else take their airplane, but the old mechanic wasn't about to argue with the group executive officer. Besides, he didn't much care who shot down Japanese aircraft with his P-38.

As soon as his wheels were off the ground, MacDonald saw the Vals. Ivey had disappeared somewhere behind and MacDonald never saw him or any other American aircraft until the battle was over. The Japanese were apparently disorganized by attacks from the 431st and 432nd Squadrons, and heading home when MacDonald raised his gear and took after a Val.

The P-38 pilot selected his first Val from one group of seven and made a head on attack until he passed overhead and swung around for a stern attack. One Val took some accurate hits right down on the surface of the water and went down to splash and explode. MacDonald started another Val smoking and the after-mission report states that this Japanese divebomber was hit by Captain Ivey and was seen to skip off the surface of the water before it crashed. Again, MacDonald did not see Ivey after takeoff and it is likely that Ivey shot

down a Val that was believed to be damaged by MacDonald only after the details of the mission were sorted out later.

Another head on pass by MacDonald got many strikes around a Val's engine cowling, but the bomber seemed to be flying away just off the surface of the water , much to the frustration of the pursuing P-38 pilot. The next Val was not so lucky when it tried to steep turn away from MacDonald's Lightning and took some hits around the cockpit before it exploded in midair.

By now MacDonald was fifty miles out to sea, but the Vals were still in range and he picked out another for attack. He made three passes and damaged his intended victim before a Zero came to the rescue. Before other P-38s arrived to drive off the Japanese fighter it had shot out MacDonald's left engine, electrical and hydraulic systems and punctured the fuel tanks.

Once more, MacDonald did not see any other American aircraft during the battle and it is only supposed that the Zero was driven off by other P-38s. MacDonald made it back to crashland at Dobodura strip number 12 without ever seeing a P-40 that was listed in mission reports as having escorted him home.

2nd Lieutenant James Molloy had witnessed MacDonald's first kill and waded into the scattered remnants of the retreating Vals. He downed one of them and badly shot up another before heading back to Dobodura.

The 433rd seemed to go wild in its first sustained attack on a Japanese formation. Lieutenant Don King followed his flight leader, Dick Kimball, down on the Vals that were no more than fifty feet off the water and watched one of them burst into flames when Kimball shot at it.

It was possible to attack Vals at will but for the competition of other 433rd P-38s that skidded or dived to get into firing position first. One Val climbed and skidded to avoid being shot down by Kimball and King, but King managed a five second deflection burst that set the bomber on fire and falling to pieces at one thousand feet.

Joe McKeon and Don Revenaugh also jumped on the hapless Vals. McKeon fired a single two-second burst to flame one Val before his guns jammed on a pass at another bomber. Revenaugh took his cue and shot that Japanese down in flames, also.

The battle was a confusion of P-38s, Zeros and Vals. The biggest problem facing the American pilots was getting into each other's way. Some P-40s from the 7th Fighter Squadron that managed to get in on the fringes of the engagement complained that P-38s attacked them as wildly as they pursued the Japanese. One 7th Squadron P-40 was apparently sent to a service squadron after being damaged by P-38 fire.

John Smith and Clarence Rieman were up together and experienced some of the tough competition for targets. One flight of P-38s was chasing what Smith identified as an Oscar somewhere below him and Smith calculated that the Japanese would try to escape by climbing and turning.

He was right. The Japanese pilot went into a steep climbing turn that gave Smith a good thirty degree deflection shot into the fighter's belly and cockpit. The plane burst into flames before it gently rolled over and went down.

Before he left the combat area Smith was joined by Lieutenant Leo Mayo of the 432nd Squadron. Mayo was probably one of the last of the 432nd in the battle area and Smith watched him shoot down another Japanese aircraft identified as an Oscar. (It is slightly possible that some Japanese Army Air Force participation was made in the operation, but more likely the tangle of aircraft that day simply led to misidentification).

The 432nd scrambled in three waves between 8:10 and 8:25. Major Tompkins led off Red and White Flights, which included Chris Bartlett and Leo Mayo, at ten minutes after eight. John Loisel led off Blue and Green Flights five minutes later and these flights included a number of pilots who would score during the interception.

Fred Harris and Vivian Cloud were the last two planes off the ground ten minutes after Loisel's flights were in the air. They had been delayed because they had to pick up their P-38s out of the revetments and taxi out to the runways. It would be their luck to arrive just in time to get in on the action.

Tomkins' flights were vectored out over Oro Bay. Within about twenty minutes after the scramble the P-38s were at 24,000 feet and a few miles off the bay when six Zeros were sighted to the east. Immediately Tomkins dived to the attack and missed a shot at the tail end Japanese.

The alerted Japanese started a hard dive toward Oro Bay and a P-38 formation leader who was silently reproving himself for his eagerness had an indicated airspeed of 550 mph while he tried to once again catch the last Zero. The P-38 fired all the way until Tomkins passed the Zero and Leo Mayo in the second flight saw the enemy plane blow up.

The two flights under John Loisel came along in time to watch the subsequent action. Lieutenant Gene McGuire in Green Flight watched Tomkins beat four other P-38s to a Val near the water and explode it. The orange flame trailing behind the doomed bomber met its reflection in the blue water of the bay.

Loisel had been searching for the enemy from his altitude of 22,000 feet when he saw the white splashes of exploding bombs five miles out to sea nowhere near the shipping. Diving down at an indicated airspeed of 450 mph, Loisel got down to 3,000 feet as quickly as he could and found a single Val with at least two Zeros as company.

Passing up the Val to deal with the Zeros first, Loisel pulled up behind one that looked Khaki-colored in the light reflected off the sea. With a single burst of gunfire the P-38 started the Zero burning. Lieutenant Jim Farris was close on Loisel's wing and fired at the second Zero which half-rolled onto its back and almost collided with Loisel. Both Zeros fell in flames into the sea.

At this point Loisel turned his P-38 to the right to clear his tail and saw two Zeros hitting the water. He later decided that one was his and the other was shot down by Lieutenant Elliott Summer, element leader in Blue Flight.

Loisel subsequently exploded yet another Zero before he and Farris were jumped by a host of Zeros who appeared out of the base of a cloud. The two P-38s did a marvelous display of low altitude evasive actions until they broke free of their attackers and raced for home. The Zeros chased them for about three minutes and then gave up the idea of catching the faster P-38s.

Lieutenants Zach Dean and Billy Gresham were in the tail end of Loisel's two flights, but managed to get into the fight, anyway. Dean led Gresham down in a dive and found a Val – perhaps the same one overflown by Loisel and Farris – just off the water.

The P-38s were going very fast and Dean managed to fire a fairly long bust that the Val dodged until the P-38 whizzed past. Gresham also passed the Val "like a cyclone", according to Dean, but apparently damaged the bomber badly enough with a quick burst to send it cartwheeling into a churning white froth on the water.

Dean found another Val at three-hundred feet and drew up to close range before firing. The bomber pilot must have been a bit rattled because he tried a split-s and went straight into the sea.

Fred Harris and Vivian Cloud had followed Major MacDonald and Captain Ivey out over the bay and watched as they went in on the tails of the Vals. Both pilots watched MacDonald shoot down his first Val before they came into the battle, diving from 6,000 feet.

At first, Harris had some trouble releasing his drop tanks (there was probably no mission where additional fuel tanks were less useful, since most P-38s barely got to altitude before shedding the things) but dropped them in time to drive right up the tail of one Val that he sent into the sea, smoking and flaming. When he pulled up again he was in time to see both MacDonald's and Ivey's Vals crash. Another Val that Harris attacked tried to steep turn out of the way, but went out of control and crashed.

Cloud could see a terrible dogfight going on above and the P-38s probably kept the Japanese escort occupied because none of the fighters came down to bother any of his passes. One Val was shot down on Cloud's first attack, but then he was unsuccessful on six or seven subsequent passes. Some of the divebombers were hit as he could see flashes on the wings of his intended victims. Other P-38s interfered with his attacks when they swooped down and cut him off from his targets. Only one of the Val rear gunners fired back at Cloud's P-38 and his fire arced over the P-38 canopy and fell off to the rear.

The 431st Squadron had scrambled on time, but was directed at first to Porlock Harbor before being ordered to turn around and go ten miles northeast of Oro Bay. 1st Lieutenant Marion Kirby was leading the sixteen P-38s and was decidedly anxious about the mission. He was the operations officer of the squadron and had been in New Guinea since the early days of 1942. Up to this scramble he had only one probable to his credit and felt the pressure of leading his pilots into what seemed to be a major air battle building, from the excited call coming over the radio.

Champlin's P-38 (right side: "We Dood it", left side: "Buffalo Blitz") before it ended up badly on October 17, 1943. (Krane collection)

The 431st climbed and milled around until it reached an altitude of nearly 20,000 feet. At least Kirby could take some comfort in the abilities of the pilots who comprised his own flight. Frank Lent was already a lathered terror in the squadron

Frank Nichols by his P-38 that fared even worse on October 17, 1943 over Oro Bay. Tom McGuire claimed three Zeros, but was shot down into the sea. Any ordinary mortal would fret sharks or Japanese, but McGuire just bobbed in the ocean for about forty-five minutes worrying what Major Nichols would say!

with three kills to his credit. Bill "Pappy" Cline and Vince Elliott were young lieutenants of proven aggressiveness with one kill each. In Cline's case he would also be proven to be a good combat leader who would eventually rise to command the squadron.

Kirby found the enemy below just as the Vals were ending their attacks. He immediately dived to the attack and pulled out at seven-hundred feet directly on the tail of a Val. All the uncertainty of just a short time ago gave way to the excitement of battle as the P-38 drew into range on its prey. With one long burst of fire Kirby pulled up to about fifty feet of his victim before he broke off for another pass.

The second pass surprised Kirby when his ammunition began detonating and flying off in various directions when it left the gun barrels. Apparently, Kirby had overheated the barrels with a super-long burst and warped them. It was no real matter, however, since both Kirby and Elliott watched the Val crash in flames.

Elliott and Frank Lent also downed Vals as well as Tom McGuire and Bob Sieber who got one each before the Zeros came down to interrupt the party. Frank Monk had fallen behind the 431st formation when his right engine began cutting out. He caught up in time to see Zeros below driving up on the tails of the leading 431st P-38s. In a forty-five degree dive he managed to get on the tail of one and send it down smoking before another Zero latched onto the tail of his own P-38.

Even with a misfiring engine Monk was able to outdive the Zero behind him, but could not keep up with the rest of the 431st and had to withdraw. The seemingly ubiquitous Jim Malloy of the 433rd was around to witness Monk's Zero fall into the water.

Lieutenant Ed Czarnecki and his wingman, Flight Officer Bill Gronemeyer, saw several Zeros flying in a giant circle and decided to attack them. One Zero maneuvered onto Czarnecki's tail, but the diving speed of the P-38 again proved superior and saved its pilot. Czarnecki made several ineffectual passes at various Zeros until Gronemeyer came in from behind and set one of the Japanese fighters on fire.

One Zero got a good shot at Czarnecki from the upper rear, but missed completely, allowing the P-38 pilot to make a rolling climb to the right and end up on his back. When he rolled out he was on the tail of his Zero attacker and fired until it fell away smoking. Czarnecki was credited with another Zero that he passed from a distance of about twenty-five feet, witnessed by Lt. Elliott. But it was not the one he attacked with the fancy maneuver.

Lieutenant Paul Morriss was leading the second element of Green Flight when he noticed the Zeros – it seemed like

Popular film stars Phyllis Brooks and Una Merkel on the USO tour that visited the 475th at Dobodura late in 1943. Brooks made such films as Lady in the Dark, as well as a number of Charlie Chan mysteries, while Merkel was featured in Destry Rides Again, and the W.C. Fields classic, "The Bank Dick." (heap via Maxwell) Below: The C-54 that brought in the USO tour. (Heap via Maxwell)

about twenty of them – coming down on their tails. He immediately pulled up and looked for his wingman, Lieutenant Bob Riegle, who was nowhere in sight. One Zero made the mistake of overshooting his P-38, which opportunity Morriss quickly exploited by firing a pair of deflection shots and the Zero went down smoking right past Frank Monk before it splashed into the sea.

Frank Lent had torn into the battle with a vengeance, but every time he had a Val or Zero in his sights another flight of P-38s would roar in and steal his thunder. Finally, he saw a lone Zero heading for home just off the surface of the water. Evidently the Japanese pilot did not see Lent who was able to drive right up the Zero's tail and shoot it down. Another Zero tried to escape a number of American fighters on its tail by pulling up into a loop and roll, but Lent used the rest of his ammunition in a good deflection shot to bring it down.

Tom McGuire got into the fight with the Zeros, too. However, his windshield ahead of the gunsight glass became hopelessly fogged and he only managed to scare one off the tail of a P-40. His wingman, Lieutenant Bob Hunt, was temporarily separated from him when McGuire decided to chase three Zeros above him about a hundred miles out to sea. Hunt managed to catch the remarkably intrepid McGuire who had drawn in close behind the trailing Zero and was furious because his guns were out of ammunition!

Post Mortem

Paul Morriss flew back to base with a vibrating engine. He landed safely and was relieved to find that Bob Riegle had made a safe landing at number 4 strip. Apparently, one of the Zeros had scored a cannon hit directly into Riegle's cockpit and wounded him slightly. Knocked out of the fight, Morriss's wingman had made it back alone for some first aid at Horanda.

The most serious loss to the 475th was 431st P-38 number 112, piloted by Lieutenant William Sieber. After he had downed one Val Sieber had his wind up and went after a lone Zero that was running out to sea. Just as he was drawing up into range Sieber was puzzled when the Zero seemed to turn around for a fight. A quick look around confirmed the P-38 pilot's worst fear when other Zeros slipped in behind to cut him off from the shore and his comrades.

It was an example of what was preached by the old hands of the group when they warned about getting caught low, slow and alone. Sieber was in a trap that he would not likely survive. The only thing he could do was keep turning and climbing and hope for the best.

Troops of C Battery, 102nd Coast Artillery as well as 2/ 5 Battery, Australian artillery, saw a P-38 heading for land, then seemingly stagger about ten miles out to sea and crash in Oro Bay. A few days later HMAS Benalla found a life raft with a badly sunburned but otherwise unharmed American in that part of the bay. A chastened Bob Sieber had indeed received grace in an impossible situation and returned to his squadron on October 25 a little wiser for the experience.

Only a few other American fighters received damage from the Japanese. On the other hand, Japanese naval air losses for the engagement included fourteen of the fifteen Vals and five of the Zeros including formation leader Aoki.

The 475th Group alone claimed twenty-one dive-bombers and fifteen fighters shot down. Obviously, there were some duplicate claims even with the given possibility of Japanese Army Oscars becoming involved in the action.

Nevertheless, it was a great victory for V Fighter Command and especially for the 475th Fighter Group. The 49th Fighter Group encountered a number of Japanese formations and claimed eighteen shot down. If the Americans overestimated the scope of the victory they were still correct in assuming that the Japanese were badly hurt and probably growing more desperate.

The evening of October 15, 1943 was a joyous one for the crews of the 475th Fighter Group. Lieutenant Virgil Hagan was gratified that his squadron, the 433rd, had claimed thirteen kills, but he confided to his tentmates that he could hardly believe that he was bold enough to down an Oscar and he wondered aloud about future missions.

Three new aces were added to the 475th's roster after the October 15 scramble. Frank Lent's triple gave him six confirmed kills while John Loisel tallied his fifth. Joe McKeon got his fourth for the 433rd Squadron added to one that he got in a P-39 with the 35th Fighter Squadron in 1942.

Tom McGuire increased his score to ten and remained the top ace of the 475th. By this time his personality seemed to have done a complete turnaround, changing a reserved exterior into talkative bravado. His widow has claimed in recent days that she was surprised when he named his P-38 "PUDGY" after a nickname for her that he hated. Also, she maintains that the general consensus among his friends in pilot training was that he was not much more than a reasonably competent pilot.

It may have been nothing in the pressure of events that led McGuire to excel as a fighter pilot. His nervous talking on the ground and shameless chatter over the radio in combat areas led to his unpopularity over a widespread part of the Fifth Air Force, not just the 475th. It is fair to say that McGuire had his defenders, but the detractors were legion.

October 17, 1943

Captain Dennis Cooper was one person who had some degree of influence over McGuire. Perhaps it was Cooper's age or the prestige of his educational background, but Cooper could get the balky McGuire to remove the disreputable battered service cap of which he was so fond, for example. In terms of his peers or subordinates, McGuire had an iron will that chafed roughly, and when he acquired rank there were more people to be chafed.

McGuire was usually a team player according to those who defend him, but he had a stiff neck for his own way of doing things. It is perhaps a trait of those who only do one or two things very well that they resist the notion that their performance can be improved. So it was with McGuire. He was remarkably courageous in combat and flew his P-38s to the point of wreckage within a very short time, but it was difficult to point out to him how to fly combat more efficiently.

One story of vague origin serves to illustrate the McGuire personality. It seems an Australian pilot was visiting the 431st area and made an unfavorable comparison of the P-38 to the Spitfire. McGuire heard of the comments and was so incensed that he challenged his ally to a mock combat. McGuire apparently had enough rank at the time to select a brand-new P-38 and turned it into junk when it was wrung completely out in the twisting maneuvers.

So it was courage and pure drive that McGuire brought to the next scramble on October 17, 1943, as much as pilot skill. The 431st was scrambled for altitude over number 12 strip at a few minutes before 9:30 in the morning to contact the enemy again over Oro Bay.

The 432nd Squadron was just returning from a flight to Lae when the scramble order was received. The formation sighted between fifteen to twenty Zeros over Oro Bay, but shortage of fuel forced the P-38s to skirt the enemy and head for home in frustration.

Sixteen P-38s of the 431st were luckier when they engaged what must have been those same Zeros at about fifteen minutes past ten. The P-38s were around 23,000 feet and the Japanese were a little higher and coming almost straight at the American interceptors.

Kirby was leading the squadron again and lost no time in taking the enemy head on. He circled to get above the enemy formation and managed to come down on the leading flight of Zeros. His first burst missed, but he turned his P-38 to the right and got in behind another Zero that was just being shot down by Ed Czarnecki.

Heading out to sea again, Kirby found a circle of Zeros in a protective Lufbery maneuver and shot at three of the turning enemy planes as if he were firing at revolving clay ducks in a shooting gallery. Czarnecki's wingman, Vincent Elliott, saw one go down in flames at almost exactly the same time that he brought down another. Both Zeros crashed about a mile apart.

During the same time, McGuire led his White Flight directly into the teeth of the Zero formation. When the 431st dropped tanks on signal, McGuire selected one Zero at the right of the formation and opened fire until his target began smoking and rolling away to the right. McGuire followed him down to 18,000 feet firing intermittently.

Once again, Bob Hunt was McGuire's wingman. He was at a disadvantage because one of his droptanks had refused to release, but he stuck with his leader long enough to see the Zero going down with heavy smoke trailing behind. Hunt couldn't pull out of the dive until he reached 4,000 feet and lost McGuire in the process. Somewhere along the line he had managed to fire at a Zero and claimed it as probably destroyed.

McGuire had climbed back alone to 21,000 feet in time to see two Zeros attacking Kirby's Red flight from above and behind. After he fired a few bursts to drive these Zeros away, four more Zeros attacked McGuire from dead astern. He was forced to dive to 14,000 feet to lose them.

For the next few minutes McGuire would climb, fire at one Zero then be chased back into a dive by others attacking him. At least two Japanese bullets found their way into the P-38 cockpit, persuading McGuire to increase the angle of his dives.

During one of his climbs to recover altitude, McGuire saw a P-38 being pursued by seven Zeros. A Zero was just pulling into firing range when McGuire made a ninety degree deflection shot that hit its mark, causing the Japanese fighter to burst into flame.

Pressing his advantage to the limit, the dauntless P-38 pilot took on the other six Zeros and caused yet a second to burst into flame.

The other five Zeros, however, were not impressed by the American's audacity. One of them got in close behind the P-38 and began firing before it could dive away once more. Cannon shells started the right engine smoking and seemed to burst all over the fighter.

McGuire was in the middle of a firestorm. One cannon shell burst right behind him in the radio and shrapnel rang off the armor plate that protected his life. Some of the bits of exploded metal had penetrated the seat and lodged in his buttocks and right arm. A 7.7 bullet had plowed right through his wrist and gone into the instrument panel. In the midst of this emergency he had time to be fascinated at the perfect hole in his wrist that produced no pain. It was obviously time to leave. Other metal fragments had damaged the control column and the P-38 could not be pulled out of the screaming dive. Releasing the top escape hatch he prepared to exit for his life.

The dive had begun at 13,000 feet and McGuire had started to abandon his P-38 several thousand feet later. He stood up in the rush of the slipstream and was perplexed when the legs of his trousers seemed to be caught in the twisted wreckage of the cockpit and he was blinded in an inky blackness. He ripped away the oxygen mask that had wrapped itself around his eyes and disentangled himself from the cockpit in time to open his parachute. Once he was in the water he could determine that his life vest would only partially inflate and the life raft was too full of shrapnel holes to inflate at all. Land was about twenty-five miles away and the cool young fighter pilot settled back to see who would reach him first.

It was about a half-hour's wait before PT-152 showed up to take the bobbing P-38 pilot back to the PT tender USS Hilo. McGuire was even lucky enough to get a statement from Lt. George G. Westfeldt, Jr. who was aboard PT-152 and witnessed the second and third Zeros crash into the sea. The first Zero was seen to crash by two U.S. Marines, Arthur Kemp and J.B. Pruett, who watched with battlefield glasses from the beach near Buna Bay.

Danny Roberts was already leading nineteen P-38s of the 433rd Squadron on a shipping cover mission when the scramble came in from the Tenth Fighter Controller Sector. Enemy planes were plotted fifty miles east of Cape Ward Hunt at 18,000 feet and would Possum please take a look.

Possum Leader (Roberts) answered to the affirmative and took the 433rd to 20,000 feet on the way. At about 10:30 twelve dark brown/green mottled Zeros were identified coming in from the southwest. The P-38s dropped tanks and broke off into flights to attack the enemy formation.

The first flight down was led by Roberts and his wingman, John Smith. Below, the Zeros became alert to the danger coming at them and turned away to dive toward New Britain. The chase was on with Roberts and Smith diving at full speed after the retreating Japanese.

It took ten minutes but Roberts drew up behind one Zero at low altitude and shot it down. The Japanese pilot appeared to have tried to bail out, but was thrown from the fighter as it crashed. The next Zero was down to 150 feet off the water and tried in vain to skid and yaw to avoid Roberts's accurate fire. When he passed over his latest kill Roberts could see the trapped pilot trying to open the canopy as the Zero was sinking.

Meanwhile, Smith had lost contact with Roberts after he had witnessed Roberts shoot down the first Zero. He had fallen behind into the second element and joined up with Lieutenant Bob Tomberg. Smith shot down one Zero and went after another that Tomberg finally got. Another Zero was jinking violently to throw off the aim of its pursuers before its pilot decided to make a steep climb.

Smith was far enough away from the climbing Zero to gradually edge up in a climb of his own to draw within 150 yards. The Zero pilot was obviously in a panic and started to turn, giving Smith a good shot from fifty yards that exploded the Japanese fighter.

Jack Fisk was having similar luck in the chase. He was flying Lieutenant Bill Horton's wing and diving after a Zero being chased by several other P-38s when the Japanese pilot pulled up and made a perfect deflection shot. Fisk's cannon shells entered the cockpit and the Zero began burning and rolled over to spin into the sea. Fisk was amazed at the lack of aggressiveness in the enemy when another Zero presented a set-up shot and Fisk flew through large pieces of the airplane when it exploded.

Lieutenants Charles Grice and Dale Meyer chased another Zero that tried aerobatics to evade the Americans. Grice was too wary to follow the Zero when it went into a loop, but waited until it levelled out and headed for Gasmata. Grice caught up with it a few minutes later and shot off the right wing with cannon fire. The Zero fell end over end into the sea.

Meyer also claimed a Zero shot down, as did Lieutenant John Howard. All claims on this mission fore the 433rd had the benefit of gun camera evidence giving the squadron a morale-boosting score of ten kills for the loss of one pilot; 2nd Lieutenant Virgil Hagan. His question about future missions was now bitterly ironic.

When Danny Roberts landed he was met by an enthusiastic crew-chief. Ted Hanks had heard about the battle, already, and jumped up on the wing of Roberts's P-38 to congratulate him, "That makes number nine for the two that you got, sir," the exuberant sergeant exclaimed.

Roberts corrected him with a smile, "That's two more that we got, sergeant." That little verbal boost almost burst Ted Hanks's buttons.

The diary entry that Danny Roberts made that night was filled with an unspoken enthusiasm: ". . . Intercepted twelve Zeros shot down ten . . . two for me. Hagan still missing. Nick's outfit lost a boy (431st Squadron's Lieutenant Edward J. Hedrick – Author) . . . McGuire bailed out. Boys all feeling O.K."

There is no mention of this engagement in translated Japanese documents of Rabaul naval air operations. It is obvious that the 433rd probably ran into a very inexperienced Zero unit that was mauled terribly. The 475th doctrine of staying in at least element formation and attacking swiftly to break away swiftly was working with remarkable effect.

Tom McGuire was shot down because of his reckless bravery that led him to attack a superior force of Zeros for the sake of another P-38 under attack. Hedrick and Hagan simply disappeared during the combat and may have been shot down or just fallen victim to some mechanical problem. By the end of the day on October 17, 1943 the 475th Fighter Group had approximately 150 confirmed kills for the loss of ten pilots in combat; an extraordinary record for just two months of operations.

Chapter VI

RABAUL

More Dobodura Notes

Life at Dobodura was becoming more amenable during the latter days of October 1943, thanks in part to the "fat cat" flights made possible by war weary B-25s and B-26s. These stripped down bombers, or in some cases retired transport planes, were responsible for bringing in fresh meat and potatoes for the mess tables. Usually, it was the bartering or begging talents of the mess sergeant responsible for better fare when, "bully beef, dehydrated horrors and 'C', ration could no longer be swallowed."

The men and officers of the group contributed to their own recreation. One of the things created by local native labor and the carpentry of "Lineham, Dossett and Kowalski, Unincorporated" was the elaborate officers' club with a canopied entrance. Known as "Club 38 475 Dobodura Avenue", the little bistro was a source of pride for the entire group.

Enlisted men had their own recreation halls with reading rooms, indoor games and whatever pleasures could be imported via the occasional fat cat flight. There were volleyball nets strung between palm trees to help forget the war during precious off duty hours.

For the 475th Group craftsman there was an abundance of diversion in materials found around the camp area. One enterprising young artisan made a metal watchband from a "genuine part of a Jap Zero." Others hammered and augured Australian coins into rings, bracelets, necklaces and charms. Australian liaison officer, Captain Percival, retaliated by making his own bits of jewelry from American coins.

Mel Allan was adjusting to Dobodura quite nicely at least for a veteran who was turning dark brown from many months exposure to the tropical sun. One day Allan and a few of his comrades had to drive down to Port Moresby and gave a whitefaced youngster a ride. When he was asked how long he had been in New Guinea the relatively fresh troop answered that he was an old timer of at least three months. For the rest of the trip he was subjected to the jeers and howls of the two-year veterans.

Another day Allan watched oxygen bottles being unloaded from a truck at Dobodura when one of the tricky things fell off the tailgate, smashing the valve. "From there on", recalls Allan, "it was a hot rocket, firing across camp into one side of the mess hall and out the other side jumped the swimming hole and went on out into the jungle, never to be seen again!"

The New Guinea natives of the area also impressed Allan. He was amazed and slightly amused at the local practice of native women using pigs as surrogate pump starters in breastfeeding. One of the women was so adequately endowed that she was able to jump start one breast via a pig in the front while the other was slung over her shoulder to feed an infant on her back.

OPPOSITE: Clarence Wilmarth had fought a long war in the SWPA, beginning with the 40th Squadron in 1942 and ending with the 433rd in October 1943. He was a classmate of Danny Roberts (41-G) and was of a similar temprament. The two of them were good friends. (Hanks)

#149 of the 432nd Squadron. This photo was probably taken at Dobodura late in 1943. Joe Forster was perhaps already in the squadron at the end of October and was already assigned to this P-38.

Rabaul, October 23, 1943.

After several aborted attempts to escort the bombers on Rabaul raids, the fighters finally got weather that would support a reasonable effort. October 23 was the day scheduled for a major B-24 strike with the P-38s sweeping ahead to catch the Japanese fighters before they reached altitude.

John Babel led the 433rd on a gunnery practice that was interrupted by a false scramble to intercept a plot during the day on October 22. Later that night he attended the naval intelligence lecture on Japanese ship and aircraft identification. The tension was already high in anticipation of a risky sweep over Rabaul.

On the morning of the 23rd, Dennis Cooper briefed his pilots about the mechanics of the sweep. There would be sixteen planes scheduled to take off, minus those who turned back for whatever reason. Weather would be 6/10 scattered cumulus from 15 to 29,000 feet over the target with decreasing scattered cumulus towards home on the return flight.

Cooper noticed Ed Czarnecki during the briefing. The dark eyes of the young pilot sparkled back intently while he sat listening to the briefing, his very unmilitary western boots and red bandana making him look like one of Garibaldi's guerril-

las. He was one of the most politically radical of all pilots in the 431st, and Cooper admired him for supporting the system and doing his duty as a fighter pilot. This rebel from Delaware was risking his life for the sake of his country, just as his comrades in the squadron were doing.

When the briefing was finished in the early hours of the morning the pilots went out to have breakfast and were transported to the Embi number 12 strip. Marion Kirby was heading the 431st again, this time as leader of White Flight. His wingman was Lieutenant Charles Samms and White Three Lieutenant Harold Holze with Vincent Elliott bringing up the rear of the flight. Right from the beginning Kirby was grim about the mission when one of the P-38s was scratched and the 431st was reduced to fifteen fighters.

Kirby led the squadron off at a few minutes after ten. At least the skies were clear enough for rendezvous with the other two squadrons. The 432nd had taken off at the same time as the 431st and the two squadrons found each other over the water at about the same altitude.

Colonel Charles MacDonald led the 432nd Squadron at the head of Red Flight, with Howard Hedrick as his wingman. Major Tomkins was next in order as leader of Clover Blue Flight. Campbell Wilson led the third flight – Clover White –

and Captain Waldman brought up the rear with Clover Green.

Danny Roberts took off at the lead of the 433rd a few minutes after the other two squadrons, but managed to get above and even slightly ahead of them before arriving over the target. P-38s of the 433rd were mainly at about 28,000 feet while the other two squadrons were between 25,000 and 27,000 feet. John Smith flew as a 433rd spare to take the place of aborts in that squadron, but the impetuous and eager pilot was only too glad when he saw the P-38 of Harold Holze turn back and slipped in ahead of a somewhat surprised Vincent Elliott.

When the 433rd arrived over Keravia Bay a few minutes after noon, it circled the area for about ten minutes before a large number of Zeros and other Japanese fighters identified as Oscars were sighted far below. Danny Roberts could see the P-38s of the other two squadrons drop their tanks and sail into the enemy. For the moment there was nothing for the 433rd to do but bide its time.

The 432nd was the first to go into action even though it was the last to arrive over the combat area. Heavy flak was bursting sporadically at the 432nd's level when Colonel MacDonald sighted about a dozen enemy fighters heading northeast over Glanges Strait. He looked around to be sure and saw a larger group climbing southwest of Rabaul; the smaller bunch was a decoy and MacDonald ordered tanks dropped to attack the larger group of fighters.

Howard Hedrick followed MacDonald down and watched him take on an Oscar. When the two P-38s passed the enemy plane Hedrick was able to take a quick glance back and saw it burst into flames. There was no time to watch the wreckage crash in the flurry of action ahead.

One Japanese plane that was coming head on looked like a Tony with no fewer than six forward-firing guns blazing away. All but one of MacDonald's guns stopped firing and Hedrick had all he could do just to keep his leader in sight. Even so, he managed to fire at another Oscar that MacDonald saw falling in flames and breaking up on the way to the water below, the Japanese pilot bailing out before the thing crashed.

Hedrick watched the pale pink parachute with a six inch black border in momentary fascination before MacDonald led him into more head on passes at attacking Oscars. With only one gun firing, MacDonald had little choice but to withdraw. Even so, there were enemy planes trailing the two P-38s that were finally chased off by Campbell Wilson who brought up Clover Blue Flight to the rescue.

The first two Oscars that Wilson saw were approaching MacDonald's flight from below and to the left. Both Oscars were very darkly painted, almost black it seemed to Wilson, and broke off quickly when Clover Blue approached. Another

Photo of Colonel MacDonald taken at Nadzab in March 1944. MacDonald's resolute escort of the B-24s on October 25, 1943 was rewarded by Japanese records that confirm his solitary fighter victory over Rabaul on that date. (Via Ethell)

Oscar came up directly behind MacDonald, but the Japanese pilot changed his mind and flew away when he realized that three P-38s were coming up from his rear.

Wilson looked behind him to clear his tail and saw a P-38 to his right with its left engine smoking badly and a Japanese

433rd Squadron P-38H-5 serial 42-66854, #177 went along on several Rabaul trips including the rough October 24, 1943 mission when the pilot was Lt. Bill Horton. On that mission the 433rd got ten confirmed victories and two probables. The 80th "Headhunters" squadron claimed twelve victories on the same mission for a record over the Rabaul area.

In spite of the fact that he was deeply involved in running the 431st by October 1943, Verl Jett managed to fly at least one Rabaul mission. (Cooper)

Herb Cochran with a native friend of the kind who got along well with easygoing American pilots. Cochran scored one of the 433rd's victories over a Zero on October 24, 1943. Photo taken a few months after the Rabaul operations were over.

fighter boring in on its tail. Acting almost on reflex, Wilson turned sharply into the enemy plane and it immediately ended its attack and flew away. Two more Oscars noticed the easy meat crippled P-38 and tried to get it before Wilson interceded and forced both of them to dive away.

Yet another Oscar came in on the cripple P-38 while Wilson was dealing with the others. Fortunately, Wilson was travelling at high speed and managed to get within one to two-hundred yards for a deflection shot that must have startled the unwary pilot for the Oscar took hits in the fuselage and left wing and made a sharp stall turn to the left. The Oscar simply fell in flames toward Blanche and was observed to crash by the pilot in the P-38 that Wilson was protecting; Lieutenant Art Wenige in #135 of the 431st Squadron.

Wenige had gone in to the attack as element leader of the 431st's Blue Flight. Marion Kirby had noticed the Japanese ganging up on one of the Clover P-38s below and ordered tanks dropped to meet the enemy fighters at 22,000 feet. Kirby's White Flight, followed by Lowell Lutton's Blue

Flight, clashed headlong into the maze of whirling P-38s and Zeros.

Blue Flight was at a disadvantage from the beginning. Lowell Lutton had experienced some misfiring of his engine at higher levels, but the P-38 seemed to calm down once the Zeros were engaged. Art Wenige's wingman had been forced to turn back before the combat area was even reached, leaving Wenige alone to shoot one Zero off the tail of Lutton's wingman. Wenige did see the Zero fly off trailing smoke and claimed it as probably destroyed. A burst of flak damaged Wenige's engine and set him up for his eventual rescue by Campbell Wilson.

Kirby, by this time, had attacked one Japanese and was climbing back to altitude when he noticed two Zeros off to his left. One was already being attacked by another P-38 and Kirby fired at the other one from directly behind, then coming

Also taken a few months after the Rabaul operations were over, this photo depicts Chase Brenizer who scored a Rabaul victory on October 29, 1943.

up to very close range, getting many hits and sending his target rolling over to head for the water below. Vincent Elliott saw the fighter, identified as a Hamp, crash.

John Smith was happy about his choice to join the 431st for the mission when he and Elliott waded into the fight. The first Zero that Smith attacked got away cleanly, but Elliott made a head on pass at one that was attacking Kirby and saw an explosion in the left wing near the fuselage. Smith observed Elliott's kill going down in a tight spin and shedding large pieces of its fuselage.

When he had settled down a bit and looked around for another target Smith went into a slight dive to meet another Japanese fighter head on. Elliott thought it was a Hamp and Smith identified it as an Oscar, but both Americans saw pieces of the enemy engine cowling fly off. The Japanese flew so close to Smith that a piece of loose engine cowling dug into the P-38's wing leading edge.

Elliott and Smith pulled up from their attack after watching Smith's kill going down in smoke and flame. No fewer than eight Japanese fighters then jumped them and chased them for several minutes until they were able to find cloud cover. Elliott did manage to attack one more Hamp with a deflection shot that went down smoking and shedding large pieces, but had to be confirmed by oath administered later under the offices of Dennis Cooper. Elliott had noticed Art Wenige's crippled P-38 and deferred to escort it back home.

The right moment for the 433rd Squadron came when the battle was raging at its fiercest. Danny Roberts had counted more than twenty-five Japanese fighters being handled very well by the other squadrons when thirty-five more enemy planes appeared below.

Roberts ordered tanks dropped and the other flights to follow him in a dive from 25,000 feet to 20,000 feet. Only Possum White Flight got the message and followed the lead flight down into the midst of the enemy. Some of the Zeros were attacking four 431st Squadron P-38s when Roberts slipped in from behind and shot one of the Japanese fighters to pieces, ripping a wing to shreds and starting the plane burning furiously.

Quickly turning to take another Zero in a quarter head on pass, Roberts fired a long burst that immediately set the plane on fire. Slowly the Zero rolled over and a large object dropped

"Nick" Nichols was becoming less prominent in the 431st during the latter days of October and beginning of November. When the Rabaul raids ended in November he was content to turn the squadron over to Jerl Jett. (Brown)

Bob Tomberg and TSgt Steward Shive. Tomberg flew several of the Rabaul raids including the one on "Bloody Tuesday", November 2, 1943. (Tomberg via Cook)

Dobodura: during one of the Rabaul raids the right wing of 433rd P-38 #172 was chewed by a B-24 prop. (Hanks)

from the cockpit. Roberts watched while the wreckage fell from 19,000 feet, but there was no parachute.

Dale Meyer was in the flight that followed Roberts and confirmed another Zero shot down. Most of the Zeros attacked by the 433rd were able to half-roll and dive away, but the four

claimed by the squadron represented the thirty-fifth confirmation for the unit.

American fighter claims for the day over Rabaul came to twelve; Ten for the 475th and one each for the 9th Fighter Squadron and 39th Fighter Squadron. No B-24s were reported

M.F. Kirby probably gave Japanese ace Fukui a hotfoot and a dunking on November 2, 1943. Evidence suggests that he did, but there were a lot of P-38s in the area even though some of the pilots seem to confirm Kirby as the victor. (Cooper)

433rd pilot Donald Y. King, "most congenial, well-liked and respected" in the words of Ted Hanks. He was missing in action after the November 2 mission to Rabaul. (Hanks)

lost, but one P-38 was missing when noses were counted back at Dobodura. Ed Czarnecki, the fighting rebel from Delaware, was gone.

One of the pilots of the 432nd had seen a yellow rubber boat far down in the water off Wide Bay while he was heading home. Even from altitude the silvery flash of rapidly moving paddles could be seen. The smoke trail of two aircraft were still visible – one from a P-38 and the other from an Oscar marking the way down into the bay.

Apparently Czarnecki had been hit by anti-aircraft fire and went down about thirty miles south of Wide Bay. By chance, a Japanese pilot had gone down and escaped to his life raft only about a mile away. While the Japanese seemed content to sit quietly waiting for rescue, Czarnecki paddled like mad to increase the distance between them and made it to shore to negotiate the various dangers of New Britain.

For their part the Japanese had intercepted the Americans with forty-eight Zeros from the 201st and 204th Kokutai led by Lieutenant Fukuda. The appearance of JAAF fighters is a puzzle since the Japanese Army was supposed to be removing its aircraft from Rabaul at this time. Translated Japanese naval documents state that two Zeros failed to return and seven others "suffered great damage."

Some Japanese reports claimed as many as twenty-five P-38s and two B-24s shot down. There were certainly duplications of claims on both sides, but the Japanese exaggerated wildly in part because of poor radio communication (often the unsatisfactory sets were removed entirely to save weight) and the new policy that denied credit to individual pilots for enemy aircraft shot down. That policy had the effect of making every observation of enemy aircraft going down into a confirmed kill whether or not the observer fired his guns.

By late afternoon the 475th P-38s had refueled at Kiriwina and most of them were back at Dobodura by five o'clock. The weather on the way back was perfect with just a few scattered clouds near the water. The overall spirit of the pilots was high and most were eager to visit Rabaul again. During the high altitude missions there was a better chance for the P-38s to score and to survive; those same eager pilots would learn a harsher lesson during the coming low level missions.

One of the 432nd pilots had another reason to be jubilant after he returned to base. Lieutenant Henry Condon got a wire from home that he was now the proud father of a baby boy.

October 24, 1943

The next day was to be the turn of the B-25s to strike the airfields at Tobera, Vunakanau and Rapapo. It would be a low

Harry Brown's P-38 after his final victory which was scored on Oct. 24, 1943 over Rabaul. The proud crewchief in the photo is probably TSgt J.D. Barrow.

level mission with bombers and escort flying near the deck on the way in and under 4,000 feet in the target area to avoid radar detection.

Eighteen P-38s of the 433rd and nineteen of the 432nd took off at 8:30 in the morning. Major Tomkins led the 432nd and Danny Roberts led the 433rd, but before the two squadrons could reach the target three 432nd P-38s and another from the 433rd turned back with mechanical difficulties.

The 431st took off at the same time with no more than ten P-38s and one of those turned back with generator trouble. 431st pilots saw one of the B-25s heading for the water just off Buna and the thought entered more than one mind that it just may turn out to be a rough mission.

Another bad omen presented itself to the 431st when the bombers crossed the New Britain coast at the mouth of the Warangoi River, near the southern end of the Gazelle peninsula. Captain Harry Brown was leading the 431st and was pleased that the trip had been made with discipline regarding the low altitude to get under enemy radar when Japanese gun batteries along the shore began firing at the bombers. Apparently the gunners were hoping to bring down some bombers

Frank Monk in the cockpit of #118 serial 42-66746. He scored his second victory on "Bloody Tuesday" over Rabaul.

with the geysers raised by the exploding shells or at least to break up the formations and divert the B-25s.

They were unsuccessful and the bombers continued on course for Rabaul. Brown was unhappy that surprise was no longer very likely, but continued on toward the eastern end of Blanche Bay on the approach to Rabaul. He had already ordered tanks dropped and a climb to 8,000 feet in preparation for expected action.

That action was not long in coming. At exactly 11:10 while the nine P-38s of the 431st were still climbing even higher about twenty-five Zeros, Oscars and a few decidedly black-camouflaged Tonys were sighted on the approach to the target. The enemy planes were scattered from six to fourteen thousand feet and the P-38s were coming up to meet them.

One Zero got into Brown's sights long enough for an ineffective burst of fire. Another Zero was more obliging and just sat still while Brown set it on fire with a well-aimed shot. The Zero rolled over slowly and Brown's wingman, Lieuten-

Vincent T. Elliott scored a double victory over Rabaul on October 23, 1943 to become an ace. (Krane collection)

ant John Cohn, as well as the rest of the flight saw the fighter explode just before it fell through a cloud layer below.

Captain Warren Lewis was bringing up the rear as leader of Hades Blue Flight with Frank Lent as his wingman. When Brown led the squadron to the attack over Vunakanau airstrip Lewis sighted about twenty Zeros and got in behind one that went into a split-s when it started taking hits. Following his prey into the roll, Lewis got about five good bursts and saw the Zero heading straight down with one wheel extended and trailing smoke.

When Lewis heard Lent call out a pair of Zeros coming down from the rear he had to be content with a probably destroyed victim. The dive that Lewis was in was near vertical, anyway, and he had to pull out in any event.

Lent had trouble with a missing engine after he and Lewis shook off the trailing Zeros. He called his flight leader and informed him that he was leaving for home. Within a few minutes a Zero had slipped in behind Lent and was doing slow rolls while it merrily blazed away at him before it simply left without warning or reason.

If the seemingly playful Zero puzzled Lent then the bewildered P-38 pilot must have shrugged in resignation when a totally black painted Tony flew in front of his plane. Not one to be taken aback very long, Lent fired a long deflection shot from sixty to ninety degrees and saw pieces fly off the tail and black smoke coming from the engine. Lent could not follow the Tony because of the P-38's ailing engine, but learned later that 501st Bomb Squadron pilot, Lieutenant Walter Kilroy, in THUMPER, B-25D, serial 41-30071, had watched the decidedly unusual Tony crash and burn.

The 432nd and 433rd had flown close escort and as a result had passed into the densest part of the Japanese interception. The 433rd led the bombers in and the 432nd brought up the rear.

The forefront of the 433rd P-38s, that flight led by Danny Roberts, was over Tobera airfield when it was "intercepted by 40/50 Oscars, Tonys and Zekes at 5/6,000 feet . . ." Roberts was later proud to report that his squadron maintained formation throughout the encounter and divided into flights and pairs to scatter the enemy fighters.

Roberts himself was down to 3,000 feet when one Zero came down from above and offered its tail as a target. One long burst from the P-38's guns and the Zero started to burn; it crashed and exploded about halfway between Vunakanau and Tobera. Lieutenant Stanley Northrup the third and final pilot in the lead flight after his element leader, Lieutenant Rieman turned back with fuel problems saw the Zero crash before the pilot could escape.

Fred Champlin came to the rescue of Kirby and shot down his second and third aerial victories on "Bloody Tuesday." (Cooper)

According to instructions, the 433rd worked its way back to the harbor and climbed to 10,000 feet in the process, fighting like mad all the time. Roberts fired at another Zero and missed, but his wingman, Jim Molloy, shot the fighter to pieces while it hung almost motionless on its back at the top of a roll at 3,000 feet.

Don King got a Zero and an Oscar for his third victory, John Babel shot down a Zero and probably another. Charles Grice, Herb Cochran and Jack Fisk were each also credited with one victory each. Grice saw his cannon shells explode around the cockpit of the Zero from which a lucky pilot successfully parachuted.

Lieutenant Bud Wire also claimed two Zeros and in the process became the latest ace of the 433rd. He had followed Lieutenant Herb Cochran and Don King down after a Zero when another Zero got in the way and forced the leading P-38s to turn in its direction. Wire took the opportunity to fire a deflection shot at the original Zero and sent it down trailing flames and large pieces of fuselage.

Wire looked around and saw King chasing a Zero with another one on his tail. When Wire's P-38 approached the second Zero, the Japanese pilot must have thought better of it and began to roll off to the right. It was a perfect set up for Wire and he fired a long burst that exploded the Zero, shearing most of its left wing off.

The 432nd Squadron was initially at a disadvantage when it crossed Wide Bay and headed for Rapopo airfield on the northwestern end of the Gazelle peninsula. The Japanese interceptors were well alerted and had enough altitude to play in the broken clouds for cover. Sixteen Clover P-38s arrived over Rabaul at 3,000 feet and their pilots could count at least fifty Japanese fighters above lining up for the attack.

Major Tomkins ordered tanks dropped, but his own were jammed under the P-38's cockpit and wouldn't release. He and his wingman pulled out of formation and his second element, Lieutenants Elliott Summer and Ray Krantz tacked onto Squareloop Harris's Clover White Flight.

Harris had decided that he could do more good for the B-25s by taking on the enemy fighters above him before they attacked first. He took his P-38s up in a high-speed climb and noticed the Japanese begin to dive on him. Clover Squadron began to close up its formation and the Japanese reacted by dispersing in every direction; the P-38 tight formation concept seemed to be justified once again.

By the time Clover reached 10,000 feet the Japanese began to engage. Harris made several good passes before he dived directly onto the tail of one Zero that took a volley of hits and went straight in. Another Zero had its canopy shot off by Harris, but spiraled into a cloud and went down as probably destroyed.

Zach Dean was following closely on Harris's wing and exploded one Zero in a spectacular shot that turned the fighter into a ball of flame. Another Zero tried to ram Dean and Lieutenant Don Willis, Clover White Three, called out a warning. The Zero missed but turned around for what was obviously another attempt.

This time Dean was ready and raked the charging Zero from one to the other with brutally accurate fire. The flying wreckage spun crazily to the ground with a pilot who was certainly dead. Someone congratulated Dean over the radio with, "Nice shooting 152 (aircraft number – author)."

While he was climbing behind Clover White, Elliott Summer saw an Oscar about two-thousand feet below and rolling slowly. Summer practically did a split-s and fired a three-second burst that tore off the left wing of the Oscar. During his pullout Summer blacked out with his fingers squeezing the triggers and his ammunition was practically gone before he recovered, but he went back to cover the bombers for the rest of the mission.

Billy Gresham and Jim Farris climbed into the battle that Hades Squadron was still fighting and Gresham downed a Zero and Farris got an Oscar. Farris had the disadvantage of a smoking engine, but shot down the Oscar, anyway, and lost the others to make a single-engine landing at Kiriwina.

Grover Gholson, leading Clover Blue, got the 432nd's sixth victory when he shot down an Oscar that crashed in the woods southwest of Tobera. The other two squadrons added twelve more confirmed kills to the total and Lieutenant Gene McGuire of the 432nd had the most badly damaged aircraft

Frank Lent scored his eight and ninth aerial kills over Rabaul. The latter on the November 2, 1943 low-level mission. He was in time to witness Kirby's Zero go down. (Cooper)

when he crashlanded at Kiriwina. Zach Dean's two kills made him an ace, as did those confirmed for Bud Wire. Danny Roberts was now only one kill behind the hospitalized Tom McGuire as top ace of the 475th with twelve victories.

A report by the Japanese 18th Army put the losses that day at two for the Army and ten "not returned" for the Navy. In addition, ten Japanese aircraft were damaged or destroyed on the ground. Lt. Fukuda again led forty-two Zeros of the 201st/204th Kokutai and claimed eighteen P-38s, One of the Japanese pilots lost during the battle was Shizuo Ishii who had been credited with twenty-nine victories during his combat career.

Major Mac Gets Through

Another high-level raid was scheduled for the next day. The 475th was able to put up only a light force by 6:45 on the morning of October 25 because of the large number of P-38s still at Kiriwina or in line for maintenance after the rigorous mission the day before. It was believed that resistance would be light because of the forty-plus Japanese interceptors claimed on October 24.

Major Charles MacDonald came down from Headquarters to fly with Clover Squadron. He led the 432nd with Zach Dean as his wingman and Lieutenant Paul Lucas leading the second element of Clover Red Flight. Lieutenant Hedrick flew the number four position. Nine other Clover P-38s made up the rest of the squadron.

By this time MacDonald had become a complete devotee of the P-38. One of his old flying comrades from the P-47-equipped 340th Fighter Squadron remembers visiting MacDonald about this time and riding with him down a 475th flightline. MacDonald stopped the jeep by one of the proud Lightnings and commented with a note of awe that the P-38 was the best fighter in the world.

Somewhere over the route one of the Clover flight lost contact and another P-38 of the remaining two flights aborted, leaving MacDonald with just seven fighters. Their task was to cover the 43rd Bomb Group behind the 90th, heading into Vunakanau airfield.

About forty-five minutes into the escort MacDonald heard the lead P-38 call out that the weather was bad and it was turning back with the bombers that it was covering. Apparently the 43rd Group did not get the message and kept right on going toward the target. The weather to the east of the target looked like solid trouble, but MacDonald decided that the bombers needed the support and took his few P-38s up to 27,500 feet to get clear of the overcast and to make his presence known to the Japanese.

The P-38s made high arcs above the remainder of the 90th Bomb Group B-24s as a gesture of moral support, which was about all such a tiny force of fighters could hope to do. Off Cape Gazelle thirty-two Zeros, Tonys (and a pair of bombers according to the Japanese documents) lined up to attack the bombers.

MacDonald led his fighters down and scattered the unenthusiastic enemy interceptors. Nevertheless, the inevitable happened and calls were heard from the bombers that Zeros were coming in from everywhere. The Zeros would make passes at the bombers until the P-38s showed up, but the Tonys kept their distance as if they were acting as spotters for antiaircraft fire.

MacDonald was kept extremely busy diving in and out of the fray, chasing off Zeros before they could press home attacks on the B-24s. Over Vunakanau, one Zero got into his sights and burst into flame from some well-aimed shots before it began the long drop to the ground. Zach Dean and at least one crew from a 43rd Bomb Group B-24 witnessed the destruction of the Zero.

The flak over the harbor was heavy, but the seven P-38s stayed in the target area for forty-five minutes. Several times the two Clover flights were attacked and maintained their formation to frustrate the Japanese fighters. All P-38s landed safely at Kiriwina by 1:30 in the afternoon.

October 29, 1943

The 475th was up again on October 26 to cover another B-25 mission to Rabaul. John Babel was frustrated when his right engine starter burned out and he experienced his first SNAFU (situation normal, all fouled up). As it turned out, the weather again refused to cooperate and the mission was scrubbed.

For the next two days the 475th pilots and crews sweated out the weather on alert for the next Rabaul raid. John Babel and Bill Horton with ground officer Captain Harry Brake had a long bull session and played records in Danny Roberts's tent until lights out to relieve the tension on the night of October 27.

The morning of October 29, 1943 was quite foggy which made the takeoff of forty-four 475th P-38s even more tense. By seven o'clock the P-38s were off in uncharacteristic ragged formations caused by various miscues to escort the B-24s once again.

This time the 431st arrived over the target, and was first into action by 12:15. However, the fortunes of battle were not with the Hades Squadron since they well succeeded in keeping the enemy off the backs of the bombers, but failed to

Ralph Bills led the 39th Fighter Squadron on the November 2, 1943 Rabaul raid. The 39th was the first V Fighter Command Squadron to claim 100 victories and scored nine more over Rabaul. (C. Bong)

confirm any Japanese shot down. Two P-38s were badly damaged; Lieutenant Paul Smith had his left engine and radio shot out with bullet holes in the canopy and gas tank and "Pappy" Cline came back with over thirty bullet holes in his plane.

Smith had the bad luck to be jumped by five Zeros and dived his smoking Lightning into a cloud to avoid them. He stuck his head out twice only to be chased back again. Finally taking a roundabout course after being hounded for half an hour he escaped and landed at Goodenough Island with about five gallons of gas left and cuts on his face and neck from flying plexiglass.

It was Possum Squadron's turn to shine again when Danny Roberts led the 433rd over Vunakanau at 12:25. The B-24s were withdrawing five minutes later when Possum came down from about 25,000 feet to hit thirty to forty Zeros, Oscars and Tonys at 18,000 feet. (Translations of Japanese documents specify seventy-five interceptors engaged the American raid.)

Roberts and his wingman, Herb Cochran, could see P-38s of the 39th and 80th Fighter Squadrons turning with Zeros all over the target area. The Japanese fighters that Roberts and Cochran chased were not very aggressive and took refuge in clouds between three and twelve-thousand feet.

One of the Zeros came back after the bombers and Roberts closed to within fifty yards and fired a twenty degree deflec-

tion shot from the tail, sending the enemy fighter down streaming flame. Roberts fired at another Zero that took heavy strikes before Cochran came up to finish it off and follow it down to watch it crash.

Lieutenant Grice followed Roberts down and missed one Zero in a steep turn to the right. Grice's wingman, Lieutenant Chase Brenizer, took off after the Zero and Grice went after another Zero on the tail of a P-38. The Zero was driven off by a long shot from Grice's guns, but tried to get on the tail of another P-38. It was an easy matter for Grice to press into close range and shoot the Zero's canopy to pieces, finally sending the flaming wreckage into a wild spin.

John Babel and Don King went headlong into the battle and were flying almost side-by-side in a pass on one Zero when another came down from above and to the left. The two P-38s turned into the attack and fired almost simultaneously, causing the Zero to emit heavy smoke before it passed directly between them and clipped King's wingtip. Babel followed the Zero down and watched it crash into the ground, but King was finally given credit for the kill.

John Smith got his sixth and last victory when he led Roberts's second element down in the bounce. Roberts and Cochran were unable to get proper lead on one Zero that Smith could easily sight from his rear position. The Zero took many hits from Smith's guns and Cochran could look back and see the flaming Zero spinning out of control. Smith's guns were burned out from the prolonged firing and his engines were misfiring, so he took refuge in a cloud and started back for home.

The 433rd was given credit for six confirmed victories while translated Japanese documents admit to no more than four Zeros lost in the air. The only American loss was Lieutenant Chris Bartlett of the 432nd.

Arthur Peregoy had led his flight with Bartlett as his wingman on a bounce through a formation of Zeros at 18,000 feet. The P-38s continued the dive at an indicated airspeed of 450 mph to lose the pursuing Zeros easily. The last time Peregoy saw Bartlett he was over Wide Bay and in no apparent trouble. By the time he reached the south shore of Wide Bay Peregoy could not find Bartlett and several passes over the area failed to locate him.

Some time after the Wide Bay area of New Britain had been secured by the Allies, Captain Floyd Bartlett of the U.S. Infantry, Christopher Bartlett's brother, visited the area to find him. A few miles inland of the point where Bartlett's P-38 had last been seen, the captain found the plane in a large swamp. It was identified by its serial number, 42-66523, the yellow bands around the propeller spinners and the large aircraft number 140 on the nose and tail.

The port engine was badly shot up, indicating that Bartlett was probably attacked after the rest of his flight had last seen him. Four of the guns were jammed, further indicating that he had limited means to defend himself and perhaps crashlanded to avert certain destruction.

Captain Bartlett found a nearby village and found out from the natives that his brother got out of the wreckage alive and unhurt and that he lived with them for about two months. Unfortunately, a German missionary who had a station in the village informed the Japanese of the pilot's presence and a patrol picked them both up.

Although the natives assured Captain Bartlett that the Japanese did not ill-treat his brother in their presence, no trace was ever found of the Clover Squadron pilot. Many prisoners at Rabaul were either summarily executed or died of the poor conditions of captivity. It is likely that Lieutenant Bartlett was one of these unfortunates.

"Squareloop" Harris is Lost

Weather prevented the scheduled Rabaul mission for October 31. The tension was growing in the sure knowledge that every day the Japanese had in respite they would prepare and rebuild their damaged air defenses. The four day delay between the October 25 and October 29 missions had given American crews an unpleasant surprise on the latter date in terms of numbers of enemy interceptors and strength of antiaircraft fire.

Captain Fred Harris decided to use the late afternoon of October 31 to test a P-38 that likely would be part of the Clover Squadron effort on the next Rabaul mission. Lieutenant Bill Ritter had been testing P-38H-1, serial 42-66595, and found that it vibrated excessively at slow speeds and medium altitudes. He landed and notified Harris who thought that the problem could have been loose drop tanks.

Gerald Johnson led the 9th Fighter Squadron on the November 2, 1943 Rabaul raid. He is standing on the extreme left and Dick Bong is directly below the number 85 on the nose of the P-38. The 9th scored 22 victories during the Rabaul raids. (Krane collection)

After he had tightened the drop tank braces Harris took the P-38 up from strip number fifteen at just one minute before four o'clock in the afternoon. That was the last thing his squadronmates ever saw of him; a troubled P-38 heading for Buna Bay in the afternoon sun.

Personnel of the 1913th Searchlight Company saw a P-38 in trouble a few minutes later about two miles off Cape Sudest. A piece of the airplane came off at about 500 feet and the P-38 went straight into the water. Some of the Clover pilots made air searches, but found nothing more than an oil slick the next day.

Bloody Tuesday

During one of the Rabaul missions probably the November 2 low level B-25 strike Ted Hanks had occasion to pour recriminations on himself. If the result had been any more disastrous he perhaps would never have come to forgive himself.

The process of preparing Possum Squadron P-38s for a combat mission is hectic to say the least. The hardworking groundcrews must attend to countless details and sometimes one of those details does not get thoroughly checked. If the detail is combined with other negative factors it could mean disaster or even death for an unsuspecting pilot.

When Roberts began his takeoff roll for the flight to Kiriwina on the first leg of the Rabaul mission the left main fuel cell cover was not secure and popped open. A beautiful but potentially dangerous column of high-octane aviation gasoline streamed up and away behind the rising P-38. Roberts and his wingman, Lewis Yarbrough, made a wide turn to the left while the rest of the squadron climbed to join them and the escaping gasoline could still be seen. Hanks watched with horror, wishing that he had made sure the cap was locked and hoping that Roberts could get by on the trip to Kiriwina.

He did get by. Immediately after takeoff Roberts noticed that there was a fuel leak and quickly found the source of trouble. He turned the fuel tank selector valve to the cross-feed position and fed both engines off the left tank. Within a short time the fuel in the open left cell was depleted and the fire danger averted. There was still enough fuel to make Kiriwina in time to continue the mission.

Possum Squadron did get into action on November 2 and 433rd records confirm a Zero for Roberts and one for Don Revenaugh, around the Lakunai airfield area. Opinion is divided on whether the confirmations should be granted, but if this was Roberts's fourteenth kill he became the leading ace of the 475th on the mission known as "Bloody Tuesday." Don King was one of the nine P-38 pilots missing after the raid to give the mission its dark sobriquet.

Two P-38s of the 80th Fighter Squadron covering a B-25 of the 501st Bomb Squadron, 345th Bomb Group. The 80th took high honors for aerial claims with 24 between October 12 and November 7, 1943.

Clover Squadron was the first into action when it arrived over Simpson Harbor just before 1:40 in the afternoon. Grover Gholson was leading the 432nd down with the B-25s at 4,000 feet over the entrance to Simpson harbor when he noticed about twenty enemy fighters 2,000 feet above. The nervous anticipation of meeting the enemy at a disadvantage was over; there they were and the Americans were at the disadvantage.

Gholson was an old hand in Southwest Pacific aerial combat. He had four confirmed kills to his credit, one of them while flying a P-39 early in the war. He had also been shot down once and walked back through the New Guinea bush to his base at Port Moresby, thus he was fully experienced in this kind of air war.

Using some of his veteran expertise he led his flight into a climb and got a little above the enemy fighters. He was able to make several ineffective passes until he saw an Oscar directly ahead and climbing. The first burst set the Oscar on fire briefly, but the second caused it to blow up.

With his wingman, Lieutenant Fostakowski, behind him Gholson got on the tail of a Zero and following it around into a turn, got in several good deflection shots. Fostakowski saw the Zero in flames after Gholson broke off his pass. Both pilots saw Lieutenant Art Peregoy hit another Zero which pulled up into a stall and then plunged in flames toward the harbor.

Gholson also witnessed the P-38 of Clover White four, flown by Lieutenant Howard Hedrick, hit another Oscar which stalled and fell off in a grotesque fiery falling leaf dive. Hedrick followed Clover White leader Leo Mayo through the battle and at one point had to avoid one Japanese fighter that seemed intent on ramming.

The fighters of White Flight were getting low on fuel and Mayo decided to head for Wide Bay and set course for home. When the flight reached Kabanga on the south of the Gazelle peninsula Mayo saw a Tony below and dived at a very high speed to get on its tail. About two-hundred yards behind his target Mayo began firing and the Tony began burning immediately.

Mayo continued pressing his attack until the Tony blew up practically in his face. Pieces of the shattered Japanese plane flew back and tore off the right wing of Mayo's P-38 which went immediately into a wild spiral. Somehow, Mayo bailed out of the madly gyrating wreckage and landed just a few yards off the beach.

There were other Japanese fighters in the air so two of the White Flight P-38s went down to cover Mayo while Hedrick stayed at higher altitude to ward off another Zero coming down. Four or five other Japanese fighters were above him, so Hedrick sparred as well as he could before trying to break away and catch the other P-38s when they headed for home.

One of the Zeros shot out Hedrick's left engine and forced him to do a classic display of single-engine performance. Tracers whipped easily past his canopy for about six or seven minutes before one of the Zeros carelessly got ahead of the P-38 and was badly hit by a burst of fire. The other Japanese soon gave up and Hedrick turned for Kiriwina to catch the rest of his flight.

Hades Squadron had the hardest luck of the day when its nine remaining P-38s led the B-25s heading for Simpson Harbor. Two or three Japanese destroyers fresh from the Empress Augusta sea battle were near the mouth of the Warangoi River when the American raiders unwittingly flew practically over their bows. Japanese destroyer leader Captain Tameichi Hara ordered his ships to fire at the American aircraft as they practically skimmed past the masts of the warships.

Art Wenige was leading Hades White Flight which had already dropped its belly tanks when he noticed heavy antiaircraft fire that he supposed was coming from ground batteries on the route to Simpson Harbor. Frank Lent was flying Hades Red Three behind Wenige's flight and saw what he thought was a battleship with a destroyer firing for all they were worth at the bombers and P-38s ahead, but could see no damage being done.

Hara believed that his gunners were shredding the American formations and, to a degree, they were. The B-25s were surprised by the unexpected heavy fire and their formations were divided. The result would later prove deadly for both the American bombers and their escorts. Japanese gunners claimed several Americans downed on the spot, but it is unlikely that any aircraft were destroyed immediately.

By this time the bombers had turned southeast and had crossed the east coast of Cape Gazelle and what seemed to Wenige to be sixty or seventy Japanese fighters came down to attack. Lutton got a ninety degree deflection shot at one Zero on the left that Wenige saw go down in flames. Then it was Wenige's turn when another Zero attacked Lutton and Wenige fired short bursts that sent the fighter down in a crazy spin into the sea, probably with a dead pilot.

It was a confused battle that fortunately had the effect of drawing the interceptors away from the B-25s to vent their fury on White Flight. Wenige again took on a Zero that was attacking Lutton and shot it down in a head on pass. Yet another Zero was coming in to attack the P-38s of White Flight and Wenige found that his guns were empty when he tried to counter this Japanese. Fortunately, Lieutenant Frank Monk was tacked onto Wenige's P-38 like a good wingman and shot the Zero down in flames.

433rd Squadron #187 landing at Dobodura around November 1943. (via Rocker)

During this time, Marion Kirby was persevering through a number of difficulties. He had got into the air late in the first place and relinquished lead of the squadron when he found that his airspeed indicator was out. Taking over Red Flight he led his P-38s into the fiery cauldron of Simpson Harbor where he was astonished to see antiaircraft gunflashes coming down on the low-flying Americans even from the slopes of the volcanoes in the harbor.

It was overcast down to the escort level of 3,000 feet with antiaircraft fire filling the air over the harbor when Wenige arrived with his flight. The P-38s jinked like mad and slipped into the clouds whenever possible to evade the black puffs of flak that seemed to be exploding everywhere. Ahead of Hades White Flight were two P-38s of Hades Red Flight that Wenige assumed were the only planes of the flight in any formation together.

Someone was calling over the radio for help, compelling Wenige to turn right and then left to see if his flight was being attacked. When he looked back to the front he could not find the P-38 he thought was being flown by Lieutenant Richardson, which left Hades Red Leader flying all alone with Zeros now coming down from above.

Wenige followed Lieutenant Lowell Lutton into a shallow dive to zigzag protectively above the bombers. The antiaircraft fire was still murderous and Hades White Flight had to take evasive action while weaving over the bombers. In the process Lieutenants Fred Champlin and Owen Giertsen became separated from the flight.

Lutton and Wenige by now down under 1,000 feet and crossing the Tobera revetments southeast of the harbor. Wenige jammed his engine controls forward to get as much speed as possible and followed Lutton in a brief strafing run. It was especially dangerous to strafe enemy airfields which were usually well-defended and Wenige breathed a sigh of relief when he cleared the area.

Over Lakunai airfield Kirby saw a Tony being shot down by a P-38. Flames were just beginning to pour back from the enemy fighter, but it was still under control. Kirby must have thought his day was complete when the doomed Tony made one last desperate try to get on his tail before it fell off into a spin and crashed below.

Flying southeast across the bay toward Rapopo Kirby was still dodging flak bursts when he saw a B-25 with its right engine on fire and running for its life from three Zeros. Although he believed that he was all alone, Kirby immediately went to the aid of the B-25. Fortunately, Frank Lent happened to catch up to Kirby and witnessed the combat.

Unknown to him at the time that he put his gunsight on the leading Zero, Kirby was drawing a bead on Lieutenant Yoshio Fukui who had arrived in Rabaul only the day before. Fukui was an aggressive Zero pilot who would be credited with eleven air victories by the end of the war.

The Japanese pilot was intent on finishing the B-25 and didn't notice Kirby sending lethal streams of bullets and cannon shells into his fighter. Somehow, the Japanese pilot survived the shock of being surprised by a P-38 and having his Zero shot away around him to parachute into the bay. He was picked up later in the day and was back in combat three days later.

Frank Lent watched Fukui's Zero fall in flames into the sea and also the B-25 go down at about the same time. Kirby had shifted his aim to the second Zero and poured hard hitting fire from his guns into the Japanese aircraft. Lent did not see what happened to this Zero, but watched Fred Champlin come up quickly to shoot down another Zero that had slipped in behind Kirby.*

One of the Zeros above came down in a diving attack on the P-38 of Lieutenant Bob Hunt. Frank Lent acted quickly and fired a few bursts that shot pieces off the Zero's engine cowling, driving the Japanese off Hunt's tail but also losing the rest of Red Flight.

It was Lent's next intention to evade the many Japanese fighters still milling overhead by getting into the overcast and working his way down toward the coast of New Ireland. Lieutenant Richardson showed up at this time and Lent could clearly see the large white number 122 on the nose of his P-38 when he drew up alongside.

On the way to New Ireland Lent sighted about six to eight Zeros making passes on two flights (6) of B-25s near the water. Most of the Zeros scattered and ran when the P-38s approached, but one made a head on pass at Lent. Frank Monk saw the brief combat from above and witnessed the Zero burn and crash when it passed Lent's P-38.

Richardson and Lent engaged a number of Zeros a few minutes later as they went after the B-25s once more. Two Zeros got on Lent's tail, but he succeeded in losing them. However, the last Lent or anybody else saw of Richardson, three Zeros were on his tail and chasing him into the distance.

Bob Hunt had his great day in combat when he claimed two zeros shot down. When he noticed that his fuel was getting low and the other P-38s were heading for the open sea, toward Kiriwina he decided to call it a day, too. Besides, he had noticed somewhere along the line that his propeller governor was out, making it unadvisable to accept combat. His P-38 landed at Kiriwina late in the afternoon and he stayed overnight until it was repaired.

Frank Lent also made for Kiriwina and landed there at about 3:30 in the afternoon with about fifteen gallons of fuel left. He was off again within two hours for Dobodura strip number twelve. Kirby also landed on Kiriwina very low on gas. He didn't know it at the time but both of the Zeros he claimed would be confirmed and he was officially an ace of the 475th.

After Fred Champlin and Owen Giertsen became separated from Hades White Flight in the clouds they were beset a number of Zeros, one of which got on Giertsen's tail. Giertsen made a break to the left, putting the Zero in a turn that crossed Champlin's path. It took only a single burst to flame the hapless Zero.

Zeros were seemingly attacking from everywhere now. Champlin dodged one by going into a cloud and emerging at the point where he was in time to give welcome assistance to Marion Kirby. Giertsen also lost the Zero on his tail, but was finally hit by one of the many deadly flak bursts in the air. He had to abandon his P-38 soon afterward near the same place that Ed Czarnecki had gone down on the October 23 mission.

Art Wenige had chased yet another Zero off the tail of Lowell Lutton and watched him join up on the tail end of his flight. Wenige knew when to leave and called Lutton to head on home. The sky was clear of enemy fighters to Wenige's delight, but, somewhere along the route to Kiriwina, Lowell Lutton disappeared. Speculation had it that he ran out of gas and quietly went into the sea. He also became an ace on this mission.

When Danny Roberts landed back at Dobodura a few minutes before five in the afternoon crewchief Ted Hanks immediately jumped up on his wing and began to apologize. Roberts gently explained that he could easily compensate for the problem and reach Kiriwina in plenty of time. Hanks said that he would never let it happen again, and Roberts smiled and said he was sure it wouldn't.

John Babel and Bill Grady took off at noon to escort the photo reconnaissance aircraft that would record the results of the mission. They spent twenty minutes circling above the flak that was still dotting the sky below and watching the ships stream out of Simpson Harbor. It was a chaotic ending to one of the bloodiest battles fought by the Fifth Air Force.

* NOTE: The identity of Fukui as Kirby's victory on November is based on a comparison of Japanese and American accounts. However, there is a slight percentage of chance that some other Zero came under Kirby's attack due to Japanese statements that Fukui first attacked the B-25 over a palm grove and that many P-38s were in the area at the time. Kirby's account is the only one that specifically mentions shooting a Zero off the tail of a B-25 and other P-38 pilots report seeing the enemy pilot take to his parachute.

NO STEP

Chapter VII

NOVEMBER-DECEMBER 1943

Rabaul Accounting

Raids on Rabaul were scheduled through November 11 when weather and other command commitments ended Fifth Air Force presence over the base. The Thirteenth Air Force continued the assault on Rabaul in conjunction with Navy and Marine air attacks, forcing the Japanese to evacuate their air forces from the base during March 1944.

The cost to V Fighter Command was in the end prohibitive. Sixteen P-38s were lost in the Rabaul fighting between October 12 and November 11, 1943. Many other Lightnings were shot to scrap, but made the flight home to at least save their pilots. Yet others were simply worn out by the demands of combat on engines and airframes.

On November 2 alone nine P-38s were lost over New Britain and at least five others received major damage. Translated Japanese documents admit the loss of eighteen Imperial Japanese Navy fighters, but no mention is made of Japanese Army Air Force fighters, which probably participated in the battle from bases on other parts of New Britain.

General Kenney remained optimistic about the effect that his forces had on reducing Japanese air power at Rabaul, but also felt the sting of his losses, especially the P-38s. During November 1943 he was obliged to reequip the 9th and 39th Fighter Squadrons with P-47s just to keep the 475th Fighter Group and the 80th Fighter Squadron of the 8th Fighter Group supplied with P-38s. For the next few months the P-38 would be a numerically subordinate fighter type in V Fighter Command.

Mel Allan had developed quite a pride in his old number 116. He is fond of recalling that one plane or another bearing the number racked up about seventeen kills while he was crewchief. A number of pilots flew into combat aboard one of the P-38s that Allan serviced and certainly contributed to the fame of 116.

At about the time of the influx of P-38s from other squadrons in November, Allan's regular pilot, John Tilley, was having sinus trouble and couldn't get to altitude. This affliction was causing him to abort missions and may have registered an unjust response from the powers that were in the 431st Squadron.

Tilley was given the job of test flying all the battered and worn P-38s coming into the 431st. Allan must have felt some sympathy for his pilot because he went along on the testing of some very doubtful aircraft. One barely flyable junker was in such bad shape that Allan wondered how it made the trip to Dobodura. He was so frustrated by the degree of maintenance required that he painted in two-inch white letters on the cowling the name "Ramblin' Wreck." Never did Allan hate so much working on one of his beloved P-38s as he did this example of the state of affairs after the Rabaul missions.

OPPOSITE: Fred Champlin got his fourth aerial victory over New Britain on December 26, 1943 but wouldn't claim his fifth until the return to the Philippines nearly a year later.

Danny Roberts Is Lost Over Alexishafen

Two new pilots came into the 432nd Squadron on October 27. Lieutenant Perry Dahl came to be known as "Pee Wee", probably with less justification than any other pilot in the USAAF. Even at the 1989 475th reunion he seemed to be a solid chunk of a man with his characteristic bright-eyed chipmunk smile. It may have been that he was named for his strikingly youthful appearance since he was still under twenty years old when he reported to Clover Squadron.

The other pilot was Lieutenant Joe Forster who, along with Dahl, would score nine confirmed air kills with the 475th. Forster would also hang up some sort of record for flying some eight-hundred miles back from a Balikpapan mission on single-engine when he had to feather one of his P-38's propellers.

After several days of orientation flights both pilots were ready for operations. Perry Dahl was scheduled for the November 9 mission to Alexishafen, a major Japanese base just above Astrolabe Bay and well north of the Huon peninsula. The young P-38 pilot would again impress his new comrades by downing a Japanese aircraft on his first combat.

Japanese construction of roads and bridges in the Madang-Alexishafen sector suggested to Allied Intelligence that the area was being used as a staging base for enemy raids against their advance lines. General MacArthur's plan was to skip over many of the bases that would be too costly and time-consuming to invade. Alexishafen would be bypassed, but it would be neutralized as a threat by heavy air attacks between November, 9 and November 16, 1943.

Once again, Major Charles MacDonald came from group headquarters to fly with Clover Squadron. Twelve 432nd P-38s took off at 8:15 in the morning to cover B-25s to Alexishafen. At about 10:20 they began diving from fifteen to eight thousand feet after the B-25s had come off their run and were being chased by Japanese fighters.

MacDonald made an identification pass on one single-engine fighter that he took to be a Zero. He overflew the fighter and told the rest of his flight to get it. Lieutenant John Rundell was on MacDonald's trail and obliged by shooting down the enemy plane.

That Zero was seen by MacDonald to crash in flames just before the Major himself got a slight deflection shot on another Zero that fell out of control and splashed into the sea. Other enemy fighters were leaving the B-25s to attend to the danger in their rear and the four P-38s of MacDonald's flight were eager to meet them.

Three other Zeros proved to be a handful when MacDonald decided to mix it up with them. One of the fighters got in

Danny Roberts's P-38 after he had been lost on the November 9, 1943 mission over Alexishafen.

behind his P-38 and put large holes in the right boom, wing and stabilizers. This Japanese upstart was probably the one claimed by Lieutenant Rundell at the same time.

Another enemy fighter diving for the Alexishafen strip was noticed by MacDonald, who went after the Japanese despite his damaged P-38. Closing on the tail of the enemy to minimum range, MacDonald fired a long burst that tore the plane into fiery pieces. The tattered remnants plunged directly into the strip.

While he was over the Alexishafen strip MacDonald noticed a P-38 in trouble below him. The Lightning was burning and its booms were coming apart before it finally crashed near the shore. Pressing combat demands on all sides permitted only the briefest moment of sorrow for the unlucky pilot who certainly died in the crash.

The pilot of the P-38 was most likely Lieutenant John Smith who probably pressed his luck once again and paid the price often demanded of the brave. Unfortunately, Smith was not the only Possum Squadron pilot who would not return from this mission.

Bob Tomberg and crewchief TSgt Shive by P-38 #177 "Miss Joanne", November 1943. (Tomberg via Cook)

Twelve P-38s led by Danny Roberts had taken off and rendezvoused over the usual point – ship wreckage off Gona at 8:30. Possum Squadron would provide low cover between six and eight thousand feet. When the bombers finished their run and headed east off the target, Roberts ordered tanks dropped to engage the pursuing enemy fighters.

Four Squadrons of 38th Bomb Group B-25s were racing toward the south and the lead Possum flight Danny Roberts with his wingman Dale Meyer followed by William Jeakle and Bill Grady in the second element ran into fifteen to twenty Oscars and Tonys over Sek Harbor between Alexishafen and Madang. With the 432nd protecting from above and the 433rd running interference below the bombers were able to get clear of the battle area.

Somewhere along the line Jeakle had to leave the fight and left for Gasmata, leaving Grady to tack on as the third plane of Roberts's flight. Grady saw what he believed to be a Hamp being attacked by Roberts and bursting into flames about one-hundred feet above the water.

There were enemy fighters everywhere, but the consensus of opinion of returning P-38 pilots was that the Japanese were disorganized and able to make only feeble attempts to fight the American escorts. Grady saw one light-colored Oscar with beautiful dark red roundels on its wings flying along just off the water. The American was able to creep up to within one-hundred yards and fire a long burst. Startled, the Oscar pilot banked sharply and splashed his fighter into the harbor.

Looking up after his sudden victory, Grady saw Roberts and Meyer chasing another Oscar toward the airstrip and fell in behind the pursuit. Within the space of a very short time, and before he could gather in the event with any sort of comprehension, Grady could see the Oscar far ahead turn sharply to the right and both P-38s ahead of him trying to turn with it. Unfortunately, Roberts was able to cut the corner much tighter than Meyer and the P-38s collided and went down to the east of the strip.

With what must have been more reflex than conscious action Grady closed on the Oscar, but could fire only a few ineffective bursts. Certainly, the impact of what he had just witnessed must have taken hold and diverted his attention. The great Danny Roberts and his wingman lost in one horrific clash of metal.

More fortunate on this mission were Jim Ince and his wingman, Perry "Pee Wee" Dahl. Dahl had seen some of the bombs hit the runway before Ince led him away to circle part of the battle over the bay. About fifteen Oscars and P-38s were in a whirling dogfight below and. Ince led Dahl down on one Japanese that was right over the water.

Somehow, this Oscar escaped by rolling away, but Dahl caught another one in a head on pass and shot it down about a half-mile from the strip into a coconut grove. Ince made a 30-degree head on shot at another Oscar that rolled under him and was seen by Howard Hedrick to go straight into the ground from 1,000 feet.

433rd Squadron P-38 #172 photographed from the tail boom of #171 late in 1943.

475th pilots accounted for nine Japanese fighters confirmed and three others probably destroyed for the three Possum Squadron P-38s and pilots lost. Major MacDonald and Captain Jim Ince both became aces during the fighting in the Alexishafen area and the group total confirmed claims now stood at 225 aerial victories.

The bodies of Danny Roberts and John Smith were eventually recovered and finally interred in the Manila American Cemetery and Memorial. Dale Meyer was never found although Bill Grady was sure his P-38 crashed only a few hundred yards from Roberts.

The Japanese must have recovered the little notebook that Roberts often carried in his shirt pocket because they cynically sent out broadcasts that he was a prisoner, using portions of the notebook as documents.

For many years after the war Ted Hanks refused to believe that Roberts had collided with Meyer. He thought that the more likely event was that Smith was so intent on a kill that he blindly ran into his squadron commander. At last the loyal old crewchief came to the same opinion as the rest of the world; his pilot and gentle commander had fallen to a random and stupid wartime combat accident.

However, at the moment that he heard the news of Danny Roberts being killed in action there was no time for recriminations. Hanks simply walked out into the bush where no one could see him and wept bitter tears.

Dobodura Animal Life

Aside from the inch-long bugs with stinking red tails and the lightning-fast lizards that streaked through the tents like green torpedoes there were the colorful wild pigeons, parakeets and birds of paradise. The meadow at Dobodura had a meandering stream that could be crossed at its head by means of a narrow, log bridge span. personnel could venture onto the narrow jungle trails by foot or jeep over the bridge.

Captain Dick Kimball and Lieutenant Louie Longman of the 433rd used the bridge for something other than convenience of travel on at least one occasion. Perhaps they wanted to compare their skill at head on aerial combat or maybe they simply had been absorbing too much jungle juice, but the pair would line up their jeeps and play a rough game of chicken by coming full speed at each other.

Some of the enlisted men were similarly jealous of their manly domain and resented having to show respect to WAC (Women's Army Corps) officers. One 431st NCO, at least, had a way to keep the women officers in their place by suddenly dropping to one knee as if to tie a shoelace just before

Above and below: Jim Ince by his P-38 with crewchief Carroll Fithian. Ince scored his fifth victory on November 9, 1943 and his final one on November 16. (Ince)

Jack Fisk's P-38 #191 with another pilot in the cockpit. Fisk got one victory on the November 9, 1943 mission over Alexishafen.

an approaching WAC officer got into saluting range. Snickers from other nearby enlisted men in dictated approval.

One Port Moresby nurse (known as the iron angel for her combination of expertise, good looks and ability to administer discipline) is said to have scotched at least some of this sort of insubordination. During a trip to Dobodura perhaps the day

433rd line late in 1943 with Danforth Miller's #187 and Fisk's #191 closest to the camera.

that Club 38 was dedicated when a number of nurses were invited from Oro Bay or Port Moresby – she was supposedly given a rude salute by one clueless troop who was summarily marched off to a private corner and treated to a heart-to-heart, eyeball-to-eyeball and nose-to-nose chewing out in the traditional military manner.

Mel Allan had a few better experiences at Dobodura. One of his relief pilots, Lieutenant Donno Bellows, was scheduled to give Allan a ride in a modified P-38. It was possible for a smallish man to squeeze into the space behind the pilot's armor in a P-38 if all radio equipment were removed. 431st Squadron policy offered short rides in such altered aircraft for those who could fit in the space behind the pilot.

When he eased into the cramped room in the cockpit Allan was bent over with the rear plexiglass panels pressing his shoulders forward and his head practically in the cockpit with

432nd line with #162 (Dahl); #146 (Dean); #159 (Peregoy) and #165 (Ratajski) in late 1943.

Probably Danforth Miller with his #187 behind him. (Anderson)

Perry Dahl in the center of the photo. He managed to claim an enemy plane in his first encounter on November 9, 1943.

the pilot. Nevertheless, Allan was in his glory when Bellows took off into the sunshine and zoomed to great heights before he twisted the P-38 into wild maneuvers, then dived and buzzed the strip.

After about ten minutes the cramped position began to have its effect on Allan and he began to feel the first queasy pangs of nausea. He touched Bellows on the shoulder and pointed down. Bellows nodded his understanding and got on the ground in record time.

It would be the ultimate moment for Mel Allan, who loved the P-38 with a passion. He had waited for this chance and mentally kicked himself for bungling at least part of it. The only consolation was that he wouldn't have to clean out the cockpit or disgrace himself all over the Lieutenant's flightsuit.

Regrouping in November

The last Fifth Air Force Rabaul mission of November 11, 1943 brought a lull of activity, but not of planning and eagerness to get at the Japanese. Neutralization of the Bismarck Archipelago would pick up again as soon as the Fifth Air Force was replenished and poised to strike.

Army and Navy air forces in the Solomons had continued the press forward toward Rabaul by landing on Bougainville on November 1. The success of this operation would further divide Japanese forces in the northern Solomons from their supply and communication base at Rabaul.

American land forces in the Southwest Pacific were preparing for Operation DEXTERITY, an offensive that would land American troops on New Britain's Cape Gloucester and Saidor, midway between Finschhafen and Alexishafen. The timetable was set by MacArthur for early December, but the Fifth Air Force had to be restrengthened and bases on the

Huon Gulf and Markham Valley had to be completed as a preliminary to major operations.

After the November 9 mission to Alexishafen there was little local air opposition to tactical missions against Japanese positions in the Huon peninsula and the Nadzab area. Many American raids were carried out with very little loss. The only thing that seemed to be in the Japanese favor was the ever present New Guinea weather which limited American air operations.

The only threat from Japanese airpower was the fighter force at Wewak and bombers based at Hollandia and Wakde Island. Early in November there were about ninety fighters reported at Wewak, but a reconnaissance on November 15 indicated that the number of fighters had grown to over one-hundred.

Early the next day about thirty 475th P-38s took off on a fighter sweep from Lae to Wewak. They would cruise along the coast between ten and twenty thousand feet with ten P-38s of Possum Squadron flying top cover at 23,000 feet. Cloud cover made visibility poor between the stacked flights, but everything from the usual rendezvous at the old Gona wreck to the final turn inland at Wewak was made on schedule.

Major Meryl Smith had come from group headquarters to lead Clover Squadron and was over the target area for about five minutes when he saw twenty-plus Japanese fighters below. He ordered tanks dropped and led the 432nd down on a good bounce. The number two Oscar in the lead formation tried to turn to the right and ran into Smith's fire that turned the once sleek little fighter into a ball of flame.

It was a happy introduction into combat for Smith who was able to attack several of the disorganized enemy fighters. He got a ninety degree deflection shot that was off a bit, but then badly damaged another Oscar which turned under his P-38 with a shot-up left wing. Another Oscar was hit in the

Nichols had put in a long tour when he handed over the 431st Squadron to Verl Jett late in November 1943. He would continue his dynamic career until he retired as a major general.

fuselage and Smith was able to see the pilot take to his parachute from the wildly spinning wreckage.

Jim Ince followed Smith and watched his first victim go up in flames. The first Oscar that Ince attacked managed to roll out of the way. Smith's flight was now lost against the green of the jungle below, so Ince went ahead and attacked the next enemy fighters that came along.

He and Lieutenant Ferdinand Hanson both damaged one of two Oscars before it escaped by rolling away under the P-38s. Ince looked in his rearview mirror and saw Smith's flight coming up from behind, making the attacks a ten-string formation.

Fifteen enemy fighters were coming in fast from a ninety degree angle and Ince quickly climbed above them and turned to face them head on. He picked one that was coming straight on and fired until it fell away in flames.

Possum Squadron was having its own trouble finding either the other P-38 squadrons or the enemy. Bill Grady was over Wewak when he saw enemy fighters taking off below and from Boram a few miles away. From over the harbor he could see twenty to twenty-five Oscars and Tonys at 7,000 feet and climbing fast toward him.

Promptly diving to the attack, Grady and Dick Kimball picked out one Oscar that turned directly into Grady. Firing from a distance of about fifty yards, Grady hit the Oscar with

devastating fire and it rolled over to explode in flames. Bill Jeakle also scored at the same time against one of the Tonys.

The pattern of patrols and sweep continued for the next few days, but Japanese opposition was less a factor with each passing day. The strength of airpower at Wewak was to be a concern for Allied planners for the next few months to come although it was more numerically powerful than the state of its actual threat. No more 475th victories would be scored over Wewak until the end of the year.

Back to New Britain

Details of cleaning up the Huon Gulf campaign were quickly being accomplished. Nadzab and Gusap were operational along the western part of the Huon peninsula and regular bombing of the Cape Gloucester proposed landing sites began on November 19.

John Babel and Captain Jim Palmer went on leave the day after Danny Roberts was killed. It took them four days to get to Sydney through bad weather and transportation bottlenecks. Two weeks later they were back in Dobodura and Babel was grounded by the 433rd flight surgeon for a head cold he picked up on the way back from leave.

During their absence, Babel and Palmer probably already knew, Captain Warren Lewis had taken over command of the 433rd. Lewis was perhaps one of the most modest pilots in the group. When the author met him at the 1989 reunion, he practically blushed at the suggestion that he was one of the great fighter pilots of the 475th.

Lewis finally had seven confirmed victories at the end of the war, but was also credited with at least eight others probably destroyed or damaged. His mission reports reflect a pronounced emphasis on operational goals rather than personal credit for enemy aircraft destroyed.

Before he was moved to the 475th Lewis was a member of the 35th Fighter Squadron that perhaps had more elite pilots in its ranks than any other Pacific unit. Legendary Major Emmett "Cyclone" Davis commanded the 35th during its heyday that included such famed pilots as Dick West, Lynn Witt and Bill Gardner. One of the things that Warren Lewis told his new squadron was, "not to mess with those guys 'cause those P-40 pilots could wax our asses if we made 'em mad."

John Babel and Warren Lewis got along very well. Babel performed the functions of an operations officer and found that he and Lewis could keep the 433rd running efficiently. Happily, their mutual abilities and respect for one another was to reflect in the coming combat record of Possum Squadron.

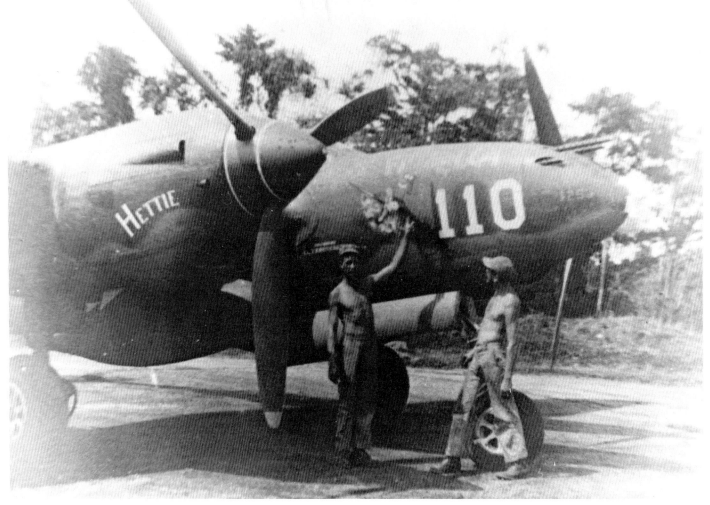

Above and below: Painting on the new number. #120 became #110 when Jett took command of the 431st. "The Woffledigit" on the left side and "Fifinella" on the right side. "Hettie" was the prerogative of the groundcrew.

Another transition in November was the passing of group command from Colonel Prentice to newly-promoted Lieutenant Colonel MacDonald on November 26, 1943. The two commanders of the 475th were together at the same time during Prentice's going home party on November 28. Prentice had been a remarkable administrative commander who gave the 475th its initial boost in training and morale.

He was not, however, a great flying commander. Some of his pilots actually expressed dread for his safety whenever he led a formation of P-38s into battle. More than once the courageous Prentice would dive into the teeth of the enemy only to oblige one or more of his pilots to chase the Japanese off his tail. In the end he apparently learned his lesson and let his pilots have all the glory.

MacDonald was a different story entirely. It would be difficult to find a better air combat leader, even in the Fifth Air Force where fine combat leaders seemed to abound. The end of the war saw MacDonald as the third-ranking fighter ace in V Fighter Command with twenty-seven confirmed victories.

The policies initiated by Prentice were maintained by MacDonald. 475th pilots and crews were encouraged throughout the Pacific war to think of themselves as "Satan's Angels", the 431st sobriquet that came to apply to the entire group by the end of 1943.

Personnel of each of the three squadrons were aware of individual personalities in the other units; it was not uncommon to hear comments like, "I mean Gene McGuire in the

Colonel MacDonald and the capable supply scrounger, Major Claude Stubbs. MacDonald took command of the 475th about the same time that Jett took over the 431st and shortly after Lewis succeeded Roberts in the 433rd.

John Tilley standing by a downed Helen bomber. Tilley's first victory was a Helen shot down over the Arawe, New Britain area on December 16, 1943. (Tilley)

432nd, not that other McGuire in the 431st", or "'Squareloop' Harris was lost this evening on a test hop", from a 433rd pilot about a 432nd pilot.

It was just as well that the change of command came as it did. The procedures begun by Prentice were sufficient to get the 475th organized and noticed by the Japanese in aerial combat (Radio Tokyo supposedly referred to the 475th pilots as the "Butchers of Rabaul.") MacDonald took command at a time when the group record was just getting up momentum and he led with the kind of fighting example and hard discipline that kept the crack group razor-sharp until the end of the war.

For several weeks the missions were routine and uneventful in the cleanup of Finschhafen and Wewak. The first real action since Alexishafen came on December 16 during a patrol of the Arawe sector of Southwest New Britain.

Tom McGuire was back on flying status and was scheduled for this mission. There was certainly no more controversial pilot in V Fighter Command; his absence since October was a breath of relief to many who despised his constant voice over the radio in combat and his return was also cheering to some who believed that, for all his faults, he was an omen of good luck with his utter fearlessness and skill in battle.

Lieutenant David Allen was leading the 431st when it took off for the patrol around 11:30 in the morning. As soon as he got the Hades Squadron on course his compass went out and he relinquished the lead to Don Bellows and took over Hades Green Flight. Within a short time Bellows had to abort the mission and Tom McGuire took the lead.

Hades Squadron was reduced from sixteen to nine P-38s when it arrived over the target area. At forty-five minutes past one in the afternoon the controller gave McGuire a plot of enemy aircraft coming in his direction from the northwest. Seven Betty* bombers and swarms of Zeros were sighted just above Cape Bushing to the north of Arawe.

The calls were coming in from various members of the Hades flights, but for some reason McGuire was unable to locate the enemy. His frustration at not finding the Japanese aircraft was increased when other P-38s took the initiative to attack.

Lieutenant Carl Houseworth led his own Hades White Flight down from about 19,000 feet with David Allen and Frank Lent of Hades Green Flight following him. Lent came down hard at the third Betty on the right side of the formation and saw pieces of canopy come off in the head on pass. Frank Monk and Lieutenant Chris Herman saw the Betty begin to break up and trail smoke.

---

* NOTE: Eventually determined to be Nakajima Ki-49 "Helen" bombers, but listed here as Bettys in accordance with 475th histories.

Glenn Jeakle in P-38J-15, serial 42-104302 later in the war. Jeakle was rammed by another P-38 on December 18, 1943 and spent ten days surviving in the New Britain jungle until he was picked up by a PBY flying boat. (Krane collection)

Allen with John Tilley on his wing made a pass from the same angle as Lent. The Betty that Allen fired upon started to smoke and turn fiercely, but the American pilot was too busy dodging the line of tracers coming from the bomber's waist position to see what was happening to his target. Lieutenant Bill O'Brien, who was coming in to attack right behind Allen's lead element watched the Betty burst into bright flames and fall like a blazing rag.

John Tilley shot at another Betty and stayed on Allen's wing until he cleared the defensive fire zones of the Japanese bomber formation. Somewhere below the now disorganized bombers the two P-38s ran into large numbers of Zeros, and Tilley tucked in close on his leader's wing until the enemy fighters were left behind.

When they were in the clear Allen and Tilley scooted back up to 20,000 feet and resumed the attack on the Bettys. One

Zero passed between the two P-38s and the startled Americans broke off their attack to once again shake off the enemy which was now coming down from above.

Once again the pair climbed for position above the bombers which by this time were flying raggedly in a nondescript formation. One of the Bettys was chosen for a stern attack and obviously became alerted when it raced for a cloudbank. Tilley started firing at about five-hundred yards and ran right up the tail of his target when he noticed that all the gunners had stopped firing. The Betty's left engine burst into flame and Tilley dived away when it was certain that the bomber was doomed.

Tom McGuire had sighted the enemy in time to follow Lent and witness his bomber kill go down. However, he was much too late to get to the surviving Bettys before they escaped into cloud. Even when he did manage to engage a straggling Zero the best he could do was inflict some damage on it.

Possum Squadron was having better luck when Warren Lewis led it to the retreating bombers. The sixteen P-38s of the squadron were at 13,000 feet when a Zero was sighted in a hole in the overcast above. As he was climbing Lewis spotted two Bettys flying side by side toward Talasea on the north coast. He picked the one on the left and closed in to fire a burst that turned the bomber into a ball of flame.

Lieutenant Charles Grice was leading the last Possum Flight and saw Lewis down his Betty and a moment later saw John Babel follow him in and shoot down the second in virtually a mirror image. Grice followed the lead flight up through the clouds and ran directly into Zeros above.

There was only one thing to do and Grice went back into the milky gray mass. When he emerged below he saw another Betty being attacked by a P-38. The apparently undamaged Betty made a slight right turn in his direction and Grice was able to send his cannon shells and tracers into the rear gunner's opening. Unhindered by the danger of the rear cannon the P-38 was able to spray its tracers into the left wing root, sending the bomber down in flames.

Meanwhile, Lewis had remained above the cloud layer and latched onto another Betty. Three enemy fighters in the air failed to deter Lewis when he got a good rear quarter shot and tore his second kill to pieces. Bill Jeakle claimed one of the Zeros and Clarence Rieman dispatched a Tony that he found somewhere to give Possum six kills for no losses in the first engagement led by its new commander.

Translated Japanese documents claim a somewhat differing view of the engagement for Imperial Naval air forces. The nearby landings of December 15 had aroused heavy Japanese attention and thirty fighter covering eight bombers had numbered six invasion ships sunk as well as five American aircraft shot down. The translation also admits the loss of two bombers and four fighters.

TSgt Edgar Childs of Magnolia, Arkansas painting the fourth victory flag garnered by various pilots with his P-38 #132 "The Magnolia Marauder."

Bill O'Brien's side of Child's P-38; "Cille." O'Brien's first victory was scored on the Wewak mission of December 22, 1943.

Recreation was highly valued during the Dobodura days, since precious little equipment could be tranported past Port Moresby. Volleyball was a good means of letting off steam and it required nothing more than a net, a pair of handy posts and the ball. (Brown)

Wewak Reprise

Another B-25 mission was scheduled to Wewak on December 22 and this time the P-38s had to fight against odds. Both the 475th Fighter Group and the 80th Fighter Squadron nearly all the operational P-38s in New Guinea were to cover a medium bomber strike at the Wewak complex. Fighter cover was scheduled at altitudes between 4,000 feet and no higher than about 8,000 feet. The altitude factor was disquieting enough for the aircrews, but the weather in the target area was reported to be increasingly cloudy.

Bill Ritter usually flew this P-38 and scored two victories over Arawe on December 21, 1943. Apparently, #154 was a lucky airplane considering the six flags below the cockpit.

Frank Monk got his third victory over Cape Gloucester on December 26, 1943 and duly recorded the fact on the P-38 that he inherited from Harry Brown.

Fourteen P-38s of the 431st arrived over the target at 9:15 AM and at no more than 4,000 feet. The B-25s made their run over Borum Strip and were on their way out to sea when about forty Oscars and Tonys suddenly appeared out of the clouds. Tom McGuire was leading Hades Green and got in one good pass at a pair of Oscars before he was forced to break away. He did manage to see Meryl Smith hit a Tony that went down in flames.

Don Bellows and Bill O'Brien did a little better when they each claimed an Oscar for Hades Squadron. 431st pilots watched a number of hits on the strip and saw burning vessels in the harbor before the P-38s retired from the area.

Clover Squadron was even luckier in claiming five Oscars and a Tony for no losses. Vivian Cloud became an ace during the battle when he claimed an Oscar and the Tony. Zach Dean got another pair of Oscars and Elliot Summer got an Oscar for his sixth kill. Perry Dahl continued his promising career by getting another Oscar.

Cape Gloucester

Throughout November and early December 1943 the nightmare of logistics on Huon Gulf and the Markham Valley delayed the prospective invasion of Cape Gloucester. Roadlinks between Lae and Nadzab were being completed in a haphaz-

#117 was the usual P-38 of Paul Morriss. His second victory was scored on December 26, 1943. P-38 was dubbed "Hold Everything."

#198, "Little Helen III" was Calvin Wire's P-38 late in 1943 and into 1944. It seems that the entire C Flight groundcrew turned out to get their picture taken.

#151 was apparently used by Stuart Lawhead after Noel Lundy was lost. Both #151, serial 42-67140, and Gresham's #168, serial 42-67147, were early camouflaged P-38J-5s. (via Rocker)

ard manner and the veteran 7th and 9th Australian Divisions were making only fitful progress in dislodging stubborn Japanese units in the muddy, malaria-ridden ground above Finschhafen.

As a result of the delays it was not until December 26 that the 1st Marine Division was landed at Cape Gloucester. Japanese response was vigorous and American claims totaled over sixty enemy aircraft destroyed, mostly Vals and Bettys. Most of the engagements took place during the afternoon patrol which was comprised mainly of P-38s of the 80th and 431st Squadrons; P-40s of the 35th Squadron and P-47s of the 36th, 341st and 342nd Squadrons.

Hades Squadron took off at 11:45 in the morning with a formation of fifteen P-38s to patrol the beachhead from one o'clock until about three o'clock in the afternoon. By the time the patrol was finished the squadron had accounted for eleven Japanese raiders.

Tom McGuire was leading Hades Squadron with Frank Monk on his wing. The controller ordered Hades and Copper (80th) Squadrons to 18,000 feet over the control vessel when the enemy was sighted coming in about fifteen miles to the northeast.

Hades Squadron climbed to 23,000 feet to get the edge on the Zero top cover which was coming straight on at 22,000 feet. Without hesitation the P-38s went head on into the enemy fighters, after McGuire had ordered the nearly dry external tanks dropped to flutter like leaves to the water far below.

Almost as soon as they had committed to attacking the Japanese top cover the P-38s of Hades Squadron were ordered by McGuire to break off and dive for the Vals that were even now reaching the ships of the invasion fleet. Frank Monk lost McGuire in the wild turning, but managed to find another P-38 from the 80th Squadron and joined it to fend off the Zero cover that was now screaming down behind the P-38s threatening the Vals below.

Monk skidded in on the tails of two Zeros that were closing in on the P-38s of his squadronmates. One lost burst of fire starting one Zero smoking badly. McGuire looked around long enough in his attack on a Val to notice Monk's P-38 and the burning Zero that fell all the way to the water.

Hades White Flight ran into a batch of Oscars below the Zero top cover. Verl Jett was leading Hades White and Vincent Elliott was element leader with John Tilley as his wingman. Elliott and Tilley made some hard turns to throw off the Oscars that were pressing on every side and Elliott hit one in a head on shot that sent it down burning and trailing black smoke.

Elliot saw two P-38s being chased by an audacious Zero and slipped in behind to shoot it down at close range. He and Tilley then joined the chase down on the water after the Vals.

Lieutenant Herman Zehring had trouble getting rid of his tanks after McGuire called out, "Let's get the dive bombers!", but finally lost them and everybody else in the process. He went straight down for the control vessel and managed to witness Frank Monk shoot down his Zero off Gloucester strip. When Zehring got down to about one-thousand feet off the water he slipped in behind a Val and exploded it with a short burst.

By this time Vincent Elliott and John Tilley were down to the level of the water and saw Zehring take after another Val that was flying just off the surface. Another short burst and Zehring sent this Val cartwheeling into the sea.

Elliott had tangled with fighters at 18,000 feet, but now watched while other P-38s took on the Vals. One Val that Elliott shot at actually bounced off the water and was set on fire by a burst from Fred Champlin's P-38. Elliott had time to look around and see one American destroyer burning badly and a P-47 pilot bail out before his Thunderbolt crashed into Borgen Bay.

Tom McGuire had legions of critics for the abrasive personal image that he projected, but nobody could accuse him of not taking his duty seriously. When he saw the dark green shapes of twenty-five to thirty Vals droning toward the invasion ships below he immediately turned his P-38 flights away from the Japanese top cover and led them in a dive toward the dive bombers.

Only one or two flights were with McGuire when he caught the Vals in their dive. One Val was caught at 8,000 feet by a short burst from McGuire's guns and was seen to crash by Fred Champlin who was bringing up the rear of the flight.

The next Val that McGuire attacked was right down on the water running for the protection of the shore between Borgen Bay and Cape Mensing. Ignoring the threat of Japanese fighters including Frank Monk's Zero kill that crashed nearby McGuire drove right up the tail of the Val and exploded it with a single burst.

John Cohn was following McGuire all through the pursuit and saw the Val blow up. Cohn himself had hit two Vals that

he believed went down, but no witnesses were around at the time. McGuire was firing wildly at any Val in range, but he settled down long enough to fire a deflection shot from thirty degrees, moving to dead astern and sent his third Val into the water.

By this time the other flights had made their way down to the water and took up the pursuit of the retiring Vals. The Thunderbolts of the 36th Fighter Squadron were also in on the chase and one of them competed with McGuire to get a Val that took some telling damage from McGuire's guns and headed inland. The P-47 pilot got credit for the kill when the Val struck a tree and exploded. McGuire relinquished the credit for his fourth victory only slowly and grudgingly over a period of time.

Paul Morriss got credit for one Val to confirm his second victory. Lieutenant Ormand Powell witnessed McGuire's third Val crash and shot down another dive bomber himself. Verl Jett was able to do no more than damage one of the Vals to complete the 431st scoring for the day.

With the confirmation of eleven more aerial victories the 431st Squadron was the first 475th unit with more than one-hundred kills. The other squadrons pushed the total score well past 250. Tom McGuire had resumed his intrepid ways and reclaimed the title of top scoring group ace with sixteen confirmed victories (seventeen victory marks would grace his P-38's scoreboard for some time).

On December 28, 1943 a haggard and worn Bill Jeakle was returned to the 433rd Squadron camp. The tail of his P-38 had been chewed off by his squadronmate, Lieutenant Austin Neely, during the battle around Arawe on December 18. Nobody was happier than the distraught Neely to see Jeakle back after eleven days in the jungle.

John Babel was also happily surprised to see Jeakle come back relatively unharmed since no parachute was seen to blossom from the tailless P-38. One of the few flying boats then on rescue service had picked up the fortunate Jeakle and taken him back to safety.

The last entry in John Babel's diary for 1943 waxes a bit laconic after the whirlwind days that preceded New Year's eve. Babel was overjoyed not just at Jeakle's lucky return, but at the sure prospect of ending his combat tour and returning home after many months in New Guinea:

"Dec 31 '43 Flew bomber escort to Alexishafen; no enemy activity. Red alert in PM. . . . Started celebrating New Year in our tent & ended up at Group club."

Nadzab was a sprawling base in the Markham Valley.

Chapter VIII

JANUARY-FEBRUARY 1944

J ohn Babel had already been ordered home on rotation when he began flying combat missions on January 1, 1944. He flew twice to Saidor that day and even did some strafing on the second mission. The next day he took his crew chief, SSgt George Mitchell, for a piggy back ride in their P-38 to Nadzab and did some acrobatics while they got pictures of Lae, Salamaua and Oro Bay.

The prospect of soon going home did not deter Babel from going the distance on risky combat missions. Even though Japanese air resistance was weak at best in the Huon Gulf area Babel flew through flak that damaged several B-24s on an escort at Alexishafen on January 4, and led another mission escorting B-24s to the same place on January 6.

He again led Possum Squadron to Ulingan Harbor near Wewak on January 8 during a B-25 escort and his final mission was leading the squadron to Madang as cover for B-24s on January 10. That was his last mission; the squadron gave him a royal farewell party in the evening and the next day he was on his way to Port Moresby where he did the usual sweating out of transport and by the end of the month was on his way through Australia – homeward bound!

MacDonald Scores Again

On the same day that John Babel flew his last mission, Colonel MacDonald flew a six-plane sweep to the Wewak area. He had favored the 432nd Squadron since joining the 475th and chose to lead from this squadron with two P-38s each from the 431st and 433rd comprising the rest of the formation.

Lieutenant Bill Ritter would fly as MacDonald's wingman. Ritter was known as an "eager beaver" who complemented his combat skill with a generous dose of enthusiasm and desire to become an ace. He had impressed MacDonald once before on a mission to Arawe when he claimed two Vals as a member of the Colonel's flight.

The six P-38s arrived over the Wewak target area a few minutes after eleven in the morning. They were at 15,000 feet flying through overcast and watching the silver shapes of about thirty Japanese fighters on Wewak strip and the west dispersal when MacDonald sighted some scattered Oscars and eight Tonys in a standard American flight formation at his own altitude.

Apparently the Tony formation was not expecting to find any enemy aircraft in their sky because MacDonald was able to slip behind the rearmost Japanese fighter and open fire within 500 feet. After taking a three-second burst the brown-mottled Tony flew for a moment then quickly rolled over and went down, Ritter watched it fall in flames.

Young Lieutenant Carrol R. Anderson just after his third mission in January 1944. Known as "Bob" or just "Andy" he was one of the most enthusiastic of the 433rd Squadron pilots and went on to record a good portion of the 475th history with his camera and pen. (Krane collection)

It became a whirling madhouse after the first Tony went down with Oscars coming in like mosquitos from every side. MacDonald turned his flight into each attack and the Japanese seemed to lose heart quickly, avoiding each pass. Incredibly, the Tonys flew on after losing one of their number as if they were unaware of what was happening. MacDonald was able to send one of the element leaders down into a spin that unfortunately was not observed by any of the now occupied P-38 pilots.

When the P-38s approached the Sepik River about fifteen Oscars came out of the overcast and tried to get on the tails of the P-38s. MacDonald turned his flight onto the rears of the newly arrived enemy fighters and started a tail chase that ended when the Japanese dived away. MacDonald only had time for a few bursts without definite results at a pair of Oscars before the Japanese were gone again.

There was just some moderate flak coming up from the direction of Nubia to dispute the aerial mastery of the 475th flight. The P-38s broke the circle they were flying and headed for Gusap to refuel before taking off for home.

More January Rotations

Major Frank Tomkins left for home on January 16 and Captain John Rundell followed two days later to give the 432nd a total loss of two valued and experienced pilots for the month. Captain Jim Ince took over as interim commander from the day that Tomkins left until John Loisel was able to devote his full energies as permanent Clover boss from January 22.

Marion Kirby also went home in January. The young former Texas A&M "Aggie" had been characterized as a tiger in combat by his comrades in both the 80th and 431st Fighter Squadrons. Kirby would say that he was just very good at conquering and masking the extreme fear that every sane fighter pilot feels. The truth is that he was an exceptional pilot and squadron sparkplug who would not easily be replaced.

His replacement was, however, a good one. Tom McGuire was glad to get the responsibility and power of Hades Operations Officer. He had two great ambitions; to become the top fighter ace of the war and to rise in rank. The reserved Thomas Buchanan McGuire had given way to the supremely confident

Left side of Anderson's first P-38 #194, "Virginia." (Anderson)

Right side of #194, "Margaret" with original crewchief, Sgt Howell. (Krane coll.)

Captain Dennis Cooper, now the 475th IO, reads the citations at an awards ceremony at Nadzab. Lined up behind him are Major Fernandez, Major Loisel, Lt. Paul Lucas and Lt. Art Peregoy. (Cooper)

fighter pilot who believed that he was destined to become the greatest of his kind.

One of the 431st officers had developed a reputation for imputing small injustices to the enlisted men. He would, for example, have a duplicate set of rules for officers and enlisted men during the various softball and other group games. It was a minor bit of tyranny, but the effect was grating on men who had relatively few pleasures in the New Guinea jungle.

Tom McGuire was one of those who took special notice of the unfairness of this officer. When he got to be C.O., McGuire was reported to have said, that officer would be gone in two days. When McGuire did take command of the 431st, the officer was given one day to pack.

A major fighter sweep to Wewak was mounted on January 18. Thirty-three P-38s took off between 8:36 and 8:45 in the morning. Colonel MacDonald was again leading at the head of Clover Squadron, with Hades following, Captain Jett at its head. During the flight to the target area, six Hades P-38s had to turn back, leaving the Squadron with twelve fighters by the time it met approximately forty Tonys and Oscars at fifteen minutes to eleven o'clock.

Even though the P-38s were at a disadvantage, since the enemy was about 5,000 feet above their own altitude of 19,000 feet, they immediately attacked. Jett got one brief chance at an Oscar and shot it down in flames. Jett could see another P-38 under attack by an Oscar below and decided to come down to its rescue despite an inoperable right inter-cooler. Even with performance limited in his P-38, he made a feint that happily worked to drive the Oscar off the tail of the P-38. With his right boom damaged, Jett was forced to shut down the engine and make for the safety of his own base.

Colonel MacDonald got into an even bigger scrap leading Clover Squadron below and to the right of Jett's formation. About forty enemy fighters came down on the rear of the Clover flights and MacDonald turned to the right to meet them. Fifteen P-38s and double the number of Tonys and Oscars clashed in one massive dog fight.

One Tony and at least two Oscars got some attention from MacDonald before they managed to twist their way out of his gunsights. He was able to see two other Tonys going down in flames and a P-38 in a tailspin; the pilot evidently was able to parachute within about five miles west of Wewak strip.

Frank Lent (center) and Tom McGuire (right) during similar award.

Champlin being congratulated by General Wurtsmith (Cooper)

Two P-38s were down to treetop level being chased by four Oscars. MacDonald could see that one of the Americans had a damaged engine with dense smoke pouring back, so he led the six remaining P-38s that were still with him down to the rescue. One of the Oscars tried to run from MacDonald, but was hit while he tried to outturn his P-38 pursuer. MacDonald was able to follow the turn and hit the Oscar again. Perry Dahl was behind the pursuit and watched the Oscar roll over and crash for MacDonald's tenth victory.

Lieutenant Howard Hedrick was in a similar hot situation to his November 1943 Rabaul mission. He was the element leader in Clover Blue Flight and tried to join up with MacDonald's Clover Green Flight when all the Japanese fighters he could possibly want came in from all sides.

Drifting out of the circle of P-38s long enough to fire an ineffective head on shot at an Oscar he got back on the wing of Lieutenant Dahl. Hedrick noticed that Dahl still had one belly tank attached and reasoned that he would be having a difficult time either attacking or defending himself.

Two Oscars then did slip into the circle and got on Dahl's tail. Hedrick fired at the first Oscar which managed to split-s out of range, but the second was not quite as fast and took hits in the right wing and fuselage. Dahl looked around in time to see Hedrick set the Oscar on fire and the Japanese pilot roll his fighter over and bail out.

Billy Gresham was in a trailing flight and came up on an Oscar that was on the tail of a P-38. With a 60 degree deflection shot, Gresham sent his tracers into the Oscar's engine and canopy, setting the plane on fire. Another Oscar was on the tail of Lieutenant Hannan in the P-38 ahead, but Gresham was at a bad angle, flying almost straight up at the Oscar. The two planes almost collided before Gresham got behind the Oscar and fired a short burst at the disappearing enemy.

Eager Beaver Bill Ritter saw his chance to get his third kill when a Tony presented itself as a target. Ritter led his wingman, Lieutenant John Michener, down to the chase and pressed in close on the attack. Like Leo Mayo during the

432nd award recipients Henry Condon, Bill Ritter, Elliot Summers and Perry Dahl at Nadzab. (Loisel via Krane files)

Sign at the approach to Finschhafen. For two weeks the 475th sweated out the odious possibility of being assigned to the disagreeable base.

Rabaul mission on which he was killed, Ritter pressed a bit too close and crashed into the Tony. Michener was shocked for a moment and when he sighted Ritter's P-38 it was going down in flames minus a wing. There was no telling if the pilot had escaped or not.

Lieutenant Bob Hadley claimed the sixth kill for the sweep when he shot down another Oscar. In addition to Michener, Lieutenants John Weldon and Joe Robertson of the 431st Squadron were missing in action. Robertson's P-38 was hit by a released belly tank and he bailed out to an obscure fate near Wewak. Weldon was last seen a few minutes after eleven o'clock at 15,000 feet a few miles south of Borum.

The losses were difficult to accept, but reflected a vigorous resurgence in the Wewak area – for the moment, at least. Both the 68th and 78th Sentai were back in full operation and the 63rd Sentai with its Oscars was reinforcing area defenses. The next few days would be difficult for both sides.

Clover Commander Loisel's First Kill

Captain John Loisel would have been the paradigm of a frontier cavalry officer. Tall and fair with a reticence that underlined his calculating intelligence he absorbed every bit of information about the enemy he fought and the P-38 that he used as a weapon. He easily passed his expertise on to the pilots of the squadron and they acknowledged him to be deft in the handling of a P-38 and more than capable of outflying the Japanese they were meeting in the air.

Charles MacDonald and John Loisel were similar in temperament and got along very well. MacDonald favored the 432nd as the squadron from which to lead the missions he flew and Loisel had his confidence in the Clover flights that followed him.

One of those missions was a B-24 escort to Wewak on January 23. MacDonald led the Clover flights off at 8:40 in the

The Nadzab area. (Krane Collection)

The Markham Valley with what is probably the Erap River running through the picture. (Brown)

morning. Within minutes the colonel was back after one of his drop tanks fell off in flight. Four more P-38s returned early with one sort of SNAFU or another, leaving Loisel in command of eleven P-38s.

Clover Squadron rendezvoused with the bombers over Gusap at five minutes after ten. The target area was reached a little over and hour later with Clover about two-thousand feet above and to the left of the bombers.

Loisel was watching the bombers turn out to sea before they began the bomb run when he saw Japanese planes just off the deck at Wewak. At the same time he saw a flight of enemy planes coming in to attack the lead formation and dropped his tanks to head them off before they could reach the bombers. Perry Dahl was flying Loisel's wing and noticed in the dive that 49th Fighter Group P-40s were behind the attacking Japanese and breaking up the threat to the B-24s.

Before they realized it Loisel and Dahl were all alone at about 12,000 feet with the rest of Clover Squadron high above after some other targets. Dahl followed Loisel who was after one Japanese fighter in the distance. Through one attack after another the faithful Dahl stayed close on his leader's wing until Loisel had a good chance at an Oscar racing for home at about one-thousand feet east of Cape Moem.

Loisel caught up with the Oscar and fired cannon shells and .50 caliber bullets until the spooked Japanese tried uncoordinated barrel rolls that failed to keep his Oscar from bursting into flames and crashing into the jungle near Brandi Plantation. Four P-38s were below and one of them was being closely pursued by Japanese fighters. Loisel and Dahl dived to the rescue.

Loisel was too late to save the P-38 which was smoking badly before it rolled over at five-hundred feet and crashed into the water near the shore. Breaking as quickly as he could before his own P-38 was bounced by prowling Japanese fighters Loisel turned sharply to clear his own tail.

One Oscar zoomed right past Dahl's P-38 in an effort to get at Loisel. Dahl cried out a warning over the radio and Loisel broke to the left along the shoreline. Pulling up to a thirty degree angle Dahl fired wildly to deter the Oscar pilot who was pulling in behind Loisel.

For whatever reason, the Oscar continued on after Loisel until Dahl was able to close on its tail, Flashes appeared on the silvery fuselage and wings of the Oscar, contrasting with the bright red markings. Smoke began to appear and the Oscar made a graceful arc downward until it hit the trees.*

Signs directing traffic to familiar New Guinea places from Nadzab.

---

* NOTE:One of the Japanese shot down over the Wewak area on January 23, 1944 was Lt. Col. Shigeo Nango, fifteen victory ace and vice commander of the 59th Sentai. Six other Japanese fighters were reported lost during the various actions over Wewak that day.

The Japanese seemed to disappear after Dahl's victory until the two P-38s started home. Dahl got a jolt when he looked in his mirror and saw it filled with the image of a Tony about fifty yards behind. Loisel immediately turned into the Japanese and drove him away.

The victorious Clover Squadron pilots made for Gusap to refuel then went home to Dobodura. For Loisel it was his eighth kill and first as commander of the squadron. Dahl was becoming quite a hero in Mercer Island, his hometown near Seattle, Washington. He had just turned twenty and had his third confirmed victory; there would be more in the near future.

It had not all gone the P-38's away on January 23 over Wewak. One 80th Fighter Squadron pilot was lost when he rammed a Japanese fighter and two Possum Squadron P-38s were lost in the same area. Loisel and Dahl undoubtedly witnessed one 433rd casualty in their unsuccessful attempt to save their P-38 comrade.

For the rest of January, aside from a record-breaking B-25 escort to Manus Island on January 24, 1944, the farthest northern target for Clover Squadron to date, action was routine and uneventful. A significant development in P-38 equipment began to appear in the 475th Group dispersals, however. The first P-38J models had appeared as early as December 1, 1943 and by the end of January 1944 they were arriving in natural metal finish.

Throughout February and the first weeks of March the groundcrews used buckets of solvent and scraping tools to remove the camouflage from older P-38s. There may have been some examples of camouflaged P-38s in operation as late as May 1944, but after midyear the appearance of the 475th P-38s was completely changed.

A Time for Leave

Regular breaks in the dull routine and horrible weather of New Guinea were the privilege of officers especially aircrew officers in the Fifth Air Force. For much of the war there was no official rotation policy; pilots and crews simply looked forward to endless missions. The only relief came in the form of short leave to Sydney for officers and lucky enlisted ranks.

John Loisel with crewchief TSgt Lawrence and two other pilots late in January or early February 1944. (Loisel via Krane Collection)

*Groundcrewman in Loisel's P-38, probably early February 1944.
(Loisel via Krane)*

On February 7, 1944 Perry Dahl left in company with four other Clover Squadron pilots on his first two-week leave to Sydney. The squadron Statistical Officer, Lieutenant Sam Fox, came back from leave on the same day and had a pleased smile to go with a spring in his step to indicate that the slight break was well worth the time invested.

A typical leave for some pilots would involve a shared apartment in Sydney or Brisbane which one or more of the renters would use during a leave. Australian women during the war were famed for their friendliness toward Americans and it was not difficult to have a few of the local young lasses up for a small party that would have the apartment joyously humming for hours.

The returning officer was obliged to come back with some item in short supply – usually beer or whiskey packed ingeniously into his furlough bag. In the case of pilots nice enough to ferry a P-38 back to New Guinea the supply of liquor could run into many gallons stuffed into every available gun bay, cockpit cranny and even wing attachments.

A turnaround of sorts was offered to the 475th Fighter Group when a troupe of dancers and singers from Sydney's Tivoli Theatre came up to Dobodura to entertain on February

12. Six live women – all of them whiter than most of the men remembered females from back in the direction of home could be – held the attention of a willing male audience in almost hypnotic power.

About this time things were changing for the enlisted men; at least those men in the 431st Squadron. Mel Allan had been on the island of New Guinea for almost two years with nothing more than an occasional emergency sick leave. When the visiting Surgeon General and Inspector asked how long he and the other enlisted ranks had waited for leave and were told the answer they screamed that more leave time be granted these poor soldiers.

Unfortunately for Allan, his first official leave to Australia was not to the bright lights of Sydney, but to officially sanctioned Camp Mackay, a small coastal town with little to recommend it. Men who returned from Mackay advised Allan to get a little dog as the quickest icebreaker in meeting the local women. He got a pup and found that the idea was a bad one; not anyone would give the little dog a tumble, let alone pay attention to the Yank sergeant who now had an added burden to cramp his social style.

Ed Czarnecki with an informal Bob Cline on his right. Czarnecki was picked up by submarine from New Britain in February 1944, but succumbed to a tuberculosis type disease that he got while evading on the island. (Cooper)

One of the first 433rd Squadron P-38J models taking off at Finschhafen in late February or early March 1944. (Hanks)

Photo taken by Ted hanks from a safe ditch to avoid exploding ammunition caused by a grass fire intentionally set that got out of control by a clean up crew. #170 is on the right, and is an early camouflaged J model before the J15 came along in March 1944. (Hanks)

P-38J-5,42-67282 #172 of the 433rd Squadron. Camouflaged P-38s became a thing of the past soon after natural metal P-38s began arriving by the end of January 1944.

Just as some of the officers did on their leaves Allan and some of his fellows rented an apartment for a week and hauled up burlap bags filled with Australian stout. Some of the Mackay residents were highly outraged that the Yanks were buying liquor on the black market and Allan was obliged several times to save himself from a beating by showing them that it was only the plentiful stout that he was lugging about.

Eight or ten men on leave caused some consternation with the landlady, but they had their own brand of fun in the apartment with the stout and whatever ration of cheap champagne or whiskey they could muster. Allan had managed to arrange a date with the young daughter of a mustachioed Turk citizen who made a point of demonstrating his curved and very sharp sword. Allan was the perfect gentleman who respected the young lady every time she said no.

Assault on the Admiralties

The decision to bypass Rabaul and secure strategic bases that surrounded the bastion was beginning to bear fruit by the end of February 1944. Nissan Island was invaded by South Pacific forces to the southeast of New Britain on February 15 and the Fifth Air Force began to neutralize the support facilities at Kavieng from the first of the month. Only weather prevented the effective crippling of the Kavieng area until the invasion of Nissan, but no interference from that quarter hampered the landings.

More than 6,000 Allied troops had been landed at Saidor on January 2, 1944 and by the end of February a force of the 1st Cavalry Division was on the island of Los Negros in the Admiralties. With these strategic positions in the hands of the Allies Rabaul and Kavieng became as caged tigers; still powerful but unable to threaten Japan's enemies. From the

Bill Gronemeyer always added one probable that he claimed before coming into the 475th. On February 3, 1944 he claimed a Lily bomber for his second official victory. If his probable claim had been upgraded, this scoreboard would be correct and he would have been another 475th ace by the end of the war. After his tour with the 475th was over he was killed in a P-38 accident.

SSgt Heap and Lt. Dahl on a P-38, probably in late 1943 – early 1944. (Heap via Maxwell)

Another group of 475th crewmen, this time from the 432nd before a camouflaged P-38J. The only identified man is TSgt Charles Bigelow, standing far right.

middle of February,13th Air Force bombers were able to strike Rabaul without air opposition of any kind.

Enemy Weather

475th operations were more effectively frustrated by wind, cloud and rain than by the Japanese during February 1944. On February 9, for example, the 432nd Squadron was on alert all day, but not a single P-38 got into the air because of the continuous pouring rain. Several escort missions got through to Kavieng during the middle of the month, but many others were scrubbed because of the weather.

In spite of the weather several 475th flights were dispersed to Cape Gloucester and Finschhafen around the middle of February. The Dobodura era was coming to an end for the group with the reduction of Rabaul and the sweep of ground forces northward along the New Guinea coast.

Perry Dahl returned from Sydney on February 20. He was well rested and eager as ever to get into a scrap with the Japanese. Air combat with Japanese fighters, however, was

still more than a month away. The 475th would continue with patrol and ground support missions for the most part until it moved to Nadzab late in March.

Dahl did get more action than he bargained for on February 29 when he was part of an eight plane ground support mission off the Cape Gloucester coast. Eighteen Clover Squadron P-38s took off at twenty minutes after seven in the morning and landed at Nadzab at 9:15.

The mission which included Dahl got off at about noon and crossed New Britain to attack positions on an island (probably Los Negros) where American troops were reportedly being overrun. A heavy rainstorm frustrated the mission about twenty miles from the target, surrounding the Clover flights when the weather also closed in behind.

Eight P-38s turned around and headed in the direction of home right on the deck. The leader of the formation followed the coast of New Britain to avoid becoming completely lost in the worsening storm. Before long the ceiling and visibility were down to zero. Two of the P-38s were lucky enough to break out of the weather and get back to Finschhafen, but the others were in a great deal of trouble.

This page and opposite: SSgt Heap on Elliot Summer's P-38, late 1943 – early 1944. (Heap via Maxwell)

As the remainder of the formation snaked along the coast just above the clutching waves Lieutenant Harold Howard ran out of luck. He must have been concerned about running into some high piece of ground directly in his path because he banked sharply as if to avoid it. Perry Dahl was occupied with keeping his own P-38 airborne, but was horrified when he saw Howard's aircraft catch a wingtip on a wave and cartwheel into the sea.

Although he didn't know it until later, two more of the P-38s in Dahl's flight, piloted by Lieutenants Dickey and Mann, crashlanded into the water near invasion Yellow Beach and the pilots were later picked up safely. Some lights appeared on the land and Dahl was heartened when he saw an airstrip off to his left.

He and Lieutenant Jack Hannan circled the pierced steel plank strip until they got a green light. Dahl immediately put down landing gear and flaps and plummeted for the runway. Soon he realized that the green light was not for him, but for a B-24 about to take off; the tower had not even seen the two P-38s.

When Dahl put his P-38's wheels on the runway the B-24 had just turned on its takeoff run from the opposite direction.

Four great engines were coming directly at the P-38 pilot and a nose-to-nose collision seemed imminent.

Instinctively ducking down into the cockpit, Dahl pressed the right brake and swerved to miss the bomber's fuselage. He avoided being crushed under the weight of the great monster bearing down on him, but only diverted his P-38 enough to go directly between the bomber's left wing engines. The powerful B-24 Hamilton Standard propellers were torn off by the P-38's tough hide and the Lockheed was spun around and thrown backward off the end of the runway.

Dahl opened his eyes to find that he was still in the present life in the same P-38 which was now resting in a gully near the rain-filled sea. He exited as quickly as he possibly could and found that, even though his P-38 was torn to bits, he had sustained no serious injury. Jack Hannan landed safely with nothing more than a broken nosewheel.

Rescue Postscript

On February 3, 1944 the U.S. submarine "Gato" left New Guinea and passed through the Vitiaz Straits on a rescue

One of the notorious scantily-clad nose-art drawings that drew negative comments from Charles Lindbergh. (Heap via Maxwell)

mission. PT 196 and PT 323 escorted the submarine until it cleared the area and was in open water headed for the Gazelle Peninsula. During the day the Gato ran submerged and raced on the surface at night.

At 10:15 on the night of February 4, the Gato was running on the surface when a large flying boat crossed its bows about three miles away, almost directly north of the Talasea Peninsula. The Gato dived and surfaced as soon as possible to clear the area quickly before Japanese destroyers could arrive after being alerted by the enemy flying boat.

The Japanese aircraft had probably come from a base on Talasea or Cape Hoskins so the Gato steered clear of the area and stood toward Open Bay at the southwestern part of the Gazelle Peninsula at top speed. During the predawn hours of February 5 the submarine reached Open Bay and began searching for the point designated for the rescue.

At first light the Gato submerged and began to scan the shore line by periscope. A white marker was sighted on the shore at eleven o'clock in the morning; that signal was suggested to Major Roberts, the Australian coastwatcher who

was gathering a number of shot down Allied airmen for rescue.

The Gato waited for dark and surfaced in a flooded down condition for a quick escape. All deck 20 mm and .50 caliber guns were manned and the bridge crew saw the prearranged flashlight signal. Two rubber boats were immediately launched after it was determined that no enemy activity was visible either on the land or seaward.

Lieutenant John Gilman and James Swanback of the Gato commanded the rubber boats through the surf and picked up Roberts who was bringing two Australian and two American airmen with him. The eager group made for the safety of the submarine without delay.

While Roberts and his rescued airmen were making toward the Gato some new and unexpected signals were coming from the beach. The bridge crew was startled when they read the signal as "sixty-seven more aviators" arriving for rescue. A Gato class submarine is fairly spacious compared to other types, but the prospect of that many personnel jammed into every corner of the boat was daunting.

Whatever the reality of the situation might have been, both rubber boats were returned to shore. When they got back to the Gato the little boats held only three more American pilots, much to the relief of the submarine's crew. Among those three brought aboard were none other than 431st Squadron pilots Ed Czarnecki and Owen Giertsen. Both pilots, in company with Lieutenant Carl Planck of the 9th Fighter Squadron, had been picked up by some of the few friendly natives in that part of New Britain.

Czarnecki had paddled to shore after he parachuted into the water on October 23 and Giertsen had crashlanded about eight miles off Wide Bay. Both men had avoided the patrols that snared other P-38 pilots shot down during the Rabaul raids and had made their way through the hostile New Britain forests to make rendezvous with the Gato by the narrowest of margins.

Actually, the three men were informed of the submarine evacuation by native runner and made it across mountains and other jungle obstacles in the shortest possible time. It was a forced march that meant the difference between freedom and possible death at the hands of perhaps the most pitiless Japanese soldiers in the Pacific.

All rescued personnel and coastwatcher Roberts were given the six-B treatment; bath, bandage, bread, butter, bullion and bed while the Gato made for the Vitiaz Straits and its faithful PT boat escort. The submarine entered Dreger Harbor in New Guinea and was moored to the USS Fortunas on the afternoon of February 7. Giertsen and Czarnecki were evacuated to Australia where Czarnecki was finally sent home to the United States.

Giertsen survived the war and surprised some of his old Hades Squadron comrades at 475th Group reunions; they had thought he was killed in action. Ed Czarnecki fared worse. The dark-haired young rebel from Maryland succumbed to some jungle disease he had picked up on New Britain.

Chapter IX

HOLLANDIA MARCH-APRIL 1944

Moving North

Even though combat in the air slowed during February and early March 1944, the pace of war quickened in New Guinea. There was time to consolidate the gains won during the past months and to advance Fifth Air Force bases within reach of the enemy once again.

On February 20, construction troops of the 475th gathered up their precious building materials and flew to Cape Gloucester. However, before they were barely aware of it, the 475th people were ordered to stop building on February 25 and were hustled off to another new base at Finschhafen. Over their loud complaints and cursing they left all their supplies and hard work to be confiscated by the local commander.

Some 475th P-38s and personnel from the 431st and 432nd Squadrons were flown into Finschhafen on February 26 and were joined by the Cape Gloucester party two days later. Finschhafen proved to be one of the worst possible bases for daily living as well as operations. The experience of the 475th contingent was that the area was a mixture of, "humid heat, torrential rains, violent electrical storms, mud and numerous day-time-biting dengue-bearing mosquitos." When the order came three weeks later for the move to Nadzab there was relief in the hearts of 475th men on the "muddy hillside" of Finschhafen.

The new base at Nadzab was like the prairie country of the American west. By the time the 475th arrived there were other units busily staking out the choicest locations for campsites and the group was consigned to the outskirts where only "antiaircraft outposts dwelt." Getting fresh water was one of the immediate difficulties and a well was dug up to a depth of twenty-five feet without success. The resulting hole was later described as the "deepest garbage pit in New Guinea."

Mel Allan had some strong impressions of Nadzab's surroundings: "The great, wide, dry, Markham river plain: never did see the town. The Erap River, now the main one, flowing swiftly through the camp area. Depth at its maximum: one-and-a-half to two feet. Very muddy and very swift. One day, to wash off the sweat-caked dirt, I attempted to sit down and soak a bit. Before I got on the bottom, the rush of water kept me only half submerged, and swept me downstream about forty feet. Who knows how far out in the ocean I would have been had I not struck another man, knocking him off his feet and crashing down on top of me?" "Afterwards, it was a funny incident – but I was almost as muddy after soaping off as before entering the water!"

One day Allan had the chance to get some real Japanese paper money from a young native who came walking into camp. He wanted to trade some Japanese invasion money to the Yanks for Australian currency and Allan could see that he had some genuine Japanese Yen mixed into the pile.

OPPOSITE: Champlin posing before a pristine #110. (Cooper)

127

V Fighter Command Headquarters sign at Nadzab. (Krane collection)

Several times Allan casually shifted one of the real Japanese bills into his trade pile and the wily native would quickly retrieve it. Evidently, he knew the difference between cheap script and the coin of the realm. After Allan had plunked down enough Australian money the native smiled when the Yankee sergeant took a few authentic notes. The local population had evidently learned a bit about horse trading from their American guests.

Nadzab was not all positive experience, in spite of its advantages over Finschhafen. The 475th camp was joined to the rest of the base by the John Britt bridge that cleared the Erap River by no more than a couple of feet. At least two 475th enlisted men were not quite so fortunate as Mel Allan when they were swept away and drowned in the swift, shallow water.

Finschhafen, also, had some good points as evidenced by a party thrown by Warren Lewis and his 433rd Squadron. Some nurses from a nearby hospital were enticed by the invitation of liquid refreshments and apparently stayed too long for the pleasure of their Medical Commanding Officer. He is loosely quoted as vowing, "Never again to let my nurses go to a party thrown by those bastards from the 433rd."

First Mission to Hollandia

Japanese fortunes had just about run their course in New Guinea. The great base at Wewak had become all but impotent and most Japanese air and ground elements were scurrying to reach Tadji or Hollandia far to the northwest of Wewak.

Colonel MacDonald in flight over the Markham Valley. (Gregg)

With the conquest of Los Negros and Manus Islands in the Admiralties, the supply line between New Guinea and New Britain was cut and the isolation of Rabaul was complete. Allied plans for the move up the New Guinea coast and eventual invasion of the Philippines could go forward with absolute safety on the flank. Only the desperate and seemingly futile defense of Hollandia remained as a Japanese option.

Fifth Air Force began limiting Japanese options by making a heavy bombing raid on March 30. Sixty-one B-24s were in the main bombing force covered by P-38s with increased fuel capacity. The P-38s would take off from Nadzab and stage through Gusap on the way back to augment their range during the extremely long-for that time-flight to Hollandia.

Clover Squadron put up twenty-two P-38s, some with the new fuel capacity, but four of the Lightnings had to abort because of faults in the system. At least six enemy aircraft were sighted, but only the 80th Squadron engaged a Japanese formation and claimed seven for no losses.

It was Hades Squadron's turn the next day when twenty-one P-38s took off at 7:55 in the morning to escort another B-24 strike to Hollandia. Rendezvous was accomplished with the bombers over Buriu (just north of the Sepik River and southwest of Wewak) about an hour and ten minutes later. Lieutenant Warren Cortner was flying squadron number 117 and ran out of oxygen a few minutes before reaching the target area. He was forced to turn back, leaving twenty P-38s to guard the bombers.

Twenty-five to thirty Oscars and Tonys were sighted at 10:15 flying well above the 431st altitude of 17,000 feet. The Japanese were coming in from the north and the P-38s braced to meet them about thirty miles south of Lake Sentani.

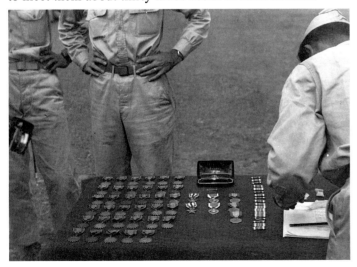

Nadzab ca. April 1944. At an awards ceremony a G.I. wrist watch is presented to MSgt Dwight N. Smith, who replaced Kenneth Greene as line chief of the 433rd, for outstanding maintenance service in the squadron. Other awards on the table include air medals, silver stars, purple hearts and bronze stars. (Hanks)

Charles Grice being decorated by General Wurtsmith. Col. MacDonald is standing nearest the camera. (Hanks)

Frank Monk was leading Hades Yellow Flight at the rear of the 431st formation and called for belly tanks to be dropped when a batch of fifteen Japanese fighters came down to the attack. The Oscars came down from the right, their fuselages gleaming silver under the mottled olive camouflage as they turned over to attack Hades Green Flight ahead.

Turning to the right to deflect the Japanese attack on his comrades, Monk got on the tail of the third Oscar and gave it a long burst of gunfire. Pieces of canopy, wing and engine flew back into the slipstream before the Oscar rolled over and went down smoking. Monk made a steep climbing turn to the right and saw the Oscar explode far below.

Monk whipped his P-38 around to the right for a few more turns to clear his flight of pursuing Japanese fighters. He saw Lieutenant Herman Zehring, his element leader, but could not find the last man, Lieutenant Robert Donald.

Lieutenant Horace Reeves, a newcomer to the 431st who was flying Monk's wing on the mission, called out enemy planes coming up from the right front. Monk was unable to locate the new threat and called out to Zehring to take over the lead of the flight.

Zehring had watched Donald roll over at the beginning of the engagement to chase a Japanese fighter below. At about

New J-15s at Nadzab preparatory to the Hollandia strikes. Captain Charles Grice, Lts. Bob Anderson, Jack Purdy and Al Pomplun were pulled off leave in Sydney to ferry planes back to New Guinea for the strikes. The ferry flight started from Eagle Farms to Townsville, then hopped the next day directly across the Coral Sea to Nadzab. (Anderson)

the same time that he saw Donald pulling up to join another flight he also witnessed Monk's victim exploding in mid-air.

When Monk called out to turn over the flight to him, Zehring had sighted ten to fifteen enemy fighters below over Humboldt Bay. He got on the tail of one Oscar which he sent crashing down into the water. A general scrap then developed and Zehring fired at a few fleeting targets until he fired a long burst at another Oscar that crashed on the shore of the bay. Hades Yellow then had to retire because of low fuel, but

Zehring smiled to himself at the confusion evident in the Japanese tactics, and the ease with which the P-38s had dealt with an evidently inexperienced and reluctant enemy interceptor force.

Hades Blue Flight was being led by Frank Lent when the Japanese came in from the right and were disrupted by Monk's Yellow Flight. Lieutenant Bob Herman and his wingman, Lieutenant Bob Crosswait, were being attacked by an Oscar and Lent raced ahead to the rescue. Crosswait managed to get

Maintenance on #150 of the 432nd Squadron at Nadzab.

Ditto on a 431st Squadron P-38.

in behind the Oscar while it concentrated on Herman and fired a few shots before Lent arrived.

The veteran ace Lent took little time to drive in on the tail of the enemy fighter and aim two bursts that immediately impressed the Oscar pilot. The brave Japanese tried to dive away, but was seen by both Lent and Crosswait to explode before he got very far.

Blue Flight was well scattered by this time and Lent managed to latch on to a formation of three P-38s at lower altitude and continued to cover the bombers. Around the target area Lent noticed the P-38 (almost certainly the one flown by Lieutenant Donald) ahead begin to trail smoke from its right

engine. Lent called out to the pilot of the ailing Lightning and warned him to head home.

The P-38 did turn around, but the white smoke that was streaming behind gradually turned to an inky black. Lent called out on the radio and urged the pilot to feather the engine without success; within a few minutes the engine burst into flames and the P-38 went into a spin.

If the troubled P-38 weren't in enough difficulty, an Oscar came down from above left to make a pass at the damaged and falling American. Lent was covering the P-38 in its death dive and stopped the Oscar with a burst of fire that set it aflame. It was too late, however, for the P-38 which began to disinte-

. . . And on #175 of the 433rd Squadron.

grate at about 5,000 feet before it exploded and hit the ground in a fiery crash.

The horror of the sight was punctuated by the crash of the Oscar which slammed in flames about a mile west of the blazing P-38. Lieutenant Merle Pearson's P-38 joined Lent's which was now circling ever higher over the crash site. Both Lent and Pearson had witnessed the two crashes and each reluctantly decided that neither American nor Japanese pilot had escaped.

Satan's Angels Over Hollandia

It was certainly a bad time for Japanese forces in the Hollandia area. The estimate for enemy aircraft destroyed by American bombers during the two missions at the end of March ran over two-hundred fighters and bombers on the ground. P-38s of the 475th Group and 80th Fighter Squadron claimed more than twenty Japanese shot down for the loss of a single P-38.

Many American individual combat reports note that the majority of Japanese interceptors on the March missions to Hollandia seemed confused, inexperienced and somewhat reluctant to press the engagements. Even with reinforcements like the 77th Sentai from the China front the Japanese were well aware of their desperate situation and responded ineffectively to the American raids.

Sometime late in February 1944 Captain Dennis Cooper turned over the duties of 431st Squadron Intelligence Officer to Lieutenant Norris Clark and moved up to 475th Group Intelligence. Cooper had been responsible in large measure for dubbing the 431st as "Satan's Angels." Now that he was in Group Headquarters the entire 475th was becoming known by the ominous sobriquet.

433rd aircraft at Nadzab before squadron colors were applied. (Anderson)

Lt. Glenn Maxwell on Nadzab after a flight. (Anderson)

"We Take Death Into the Skies" had become the official motto of the 431st for very literal reasons. By the end of March 1944 the squadron had over 115 confirmed kills (the 9th Squadron led the pack with over 180 kills for a V Fighter Command unit, while the 80th had about 170 and the 39th followed with 130-odd victories before it gave up the P-38 for P-47s in November 1943). John Tilley is generally given credit for designing the fearsome demon head logo of the 431st.

Hollandia was given a brief respite on April 1 and 2, which also gave the 475th a chance for needed maintenance on its P-38s. The breather provided by bad weather, which had always hampered operations in New Guinea and would even prove

Pilots of the 432nd ready for the Hollandia missions – left to right: Larry Roberts, Dean Olson, Jack Hannan, Bob Kimball, Bob Shuh, two in back are unidentified. In the drivers seat is Joe Forster, then Dallas Peaveyhouse, and Perry Dahl.

deadly in the near future was certainly more useful to the Americans, since the Japanese had to make major repairs to their base before the next great blow fell on Hollandia the next day.

With the added bit of time for repair of its fighters the 475th was able to put up fifty-five P-38s by a few minutes after nine in the morning. Colonel MacDonald was leading at the head of Clover Red Flight with the faithful Lieutenant Jack Hannan as his wingman once again. Henry Condon was leading Clover Green, John Loisel was Clover Blue leader with Perry Dahl as his number three and Elliot Summer had charge of Clover White Flight. Lieutenant Clifford Mann led a spare two-plane flight with Flight Officer Joe Barton as his wingman.

Clover Squadron arrived over the target area with a force of B-25s and A-20s at 11:35. When John Loisel reached the southeast edge of Lake Sentani with his Clover Blue Flight he heard the A-20 crews call out Japanese fighters over the radio. The P-38s were at an altitude of 8,000 feet and four pairs of eyes in Clover Blue searched around, behind, below and above for the enemy.

Loisel sighted one group of Japanese fighters flying in a loose string heading west. He then looked down and saw another group of fighters flying just off the deck. There seemed to be eight Oscars in the formation below in two flights; the first was a diamond-shaped flight of four followed by a similar number in something of a modified echelon or rough finger-four string.

There was no hesitation in deciding which group to attack. With a minimum of wasted movement Loisel dove on the lower group of Oscars and came in quickly on their tails. He opened fire on the leader from a range of 300 yards and then switched to the wingman, pressing in to a range of only fifty yards. Strikes appeared all over the Oscar's fuselage and wingroots. The Japanese fighter staggered and broke to the left before it fell off and crashed in the short distance to the ground below.

During the next few minutes Clover Blue was engaged with a host of Japanese interceptors. Loisel was getting mostly short bursts at fleeting targets that offered little more than difficult deflection shots before another Oscar showed its tail out front. A brief burst from Loisel's guns cause the Japanese pilot to pay for the lapse by spinning into the side of a mountain and exploding in flames. Jack Luddington was flying on Loisel's wing and witnessed the Oscar smack into the unyielding earth.

While Loisel and Luddington were busy chasing one Oscar or another, the second element of Clover Blue Flight was scoring its own victories. Perry Dahl was leading his old

Colonel MacDonald with General Wurtsmith, probably at Nadzab in March or April 1944 when MacDonald's score stood at eight.

comrade Joe Forster into the teeth of the enemy. Forster was the first to draw blood when he caught two Oscars flying together and shot both of them down near the shore of Lake Sentani.

Dahl witnessed Forster's victims crash and saw at the same time another P-38 under attack. The P-38 was flown by Lieutenant John Temple who had come over from Clover Green Flight to lend a hand with the fighting. One of the Oscars managed to get behind Temple who escaped by climbing into the clouds. When Temple emerged again from the gray silky mist he was in time to see Dahl pull up after shooting down his erstwhile assailant.

One of the Oscars then fell to Temple's guns and splashed into Sentani Lake. Dahl saw this Oscar go down and riddled another so badly that it apparently tried to land, but crashed at the end of Sentani Strip. Forster was duelling with another Oscar in a series of head on passes when he got good strikes all over the fuselage. The Japanese pilot was trying to get out while the Oscar dived toward a mountain and was actually on the wing when Forster gave the doomed plane a short burst that sent it and its pilot down to destruction.

Clover Green Flight had become separated during the initial contact, just after external tanks had been released to take on six Oscars flying in from the north at 2,000 feet. Lieutenant Henry Condon saw John Loisel shoot down his first Oscar, and then looked around to notice that only Clover Green Three, Lieutenant Laurence LeBaron, was still with him.

On the south side of Sentani Lake, Condon could see an Oscar at about five-hundred feet being pursued by six P-38s. The Oscar was doing a masterful job of turning inside the fire of his attackers because they were consistently overshooting him while he skidded and turned at about 150 miles an hour.

Condon cooly cut his throttles to get his airspeed down to that of the Oscar and slipped in behind. As soon as the American started firing the Oscar turned to the right, allowing Condon to get a two radii lead in his gunsight. Both Condon and LeBaron saw the Oscar take hits in the wings and cockpit before it began to smoke slightly and dived into the trees below.

Elliot Summer led Clover White Flight down on an Oscar attacking the A-20s. Through bursts of antiaircraft fire Sum-

Bob Tomberg, Austin Neely, Herb Cochran, Louis Longman, Bill Grady and Clarence Rieman of the 433rd Squadron early in 1944. Tomberg, Neely and Longman all went down on the infamous "Black Sunday" and only Tomberg managed to make it back. (Lewis via Cook)

Elliot Summer, Bill Ritter and Henry Condon. Summer and Condon each claimed an Oscar over Hollandia on April 3, 1944.

Probably Verl Jett's natural metal #110 in the early months of 1944. (Krane coll.)

mer got into very close range and shot the Japanese down in flames. In the ensuing engagement Summer saw a P-38 with an Oscar on its tail directly above his own Lightning. Pulling straight up and firing until he stalled, Summer convinced the certainly startled Oscar pilot to break off the attack.

Colonel MacDonald's Clover Red Flight found six Oscars approaching the A-20s on the southeast side of the lake. MacDonald had already ordered tanks dropped and led his P-38s down to split up the enemy attack. One Oscar stayed in the area long enough to become the one shot down by Elliot Summer, witnessed by MacDonald.

Jack Hannan followed MacDonald in an attack on another Oscar that had to be broken off to avoid crashing headlong into the mountains. The two P-38s circled the towering peaks until they came back toward Hollandia strip. Hannan then got in a forty-five degree shot at an Oscar that reefed in such a tight turn that it almost managed to ram its P-38 attacker before crashing in flames.

Clover Red Flight made a few more passes in the area before it flew over Humboldt Bay to rejoin the bombers. By ten minutes after the noon hour they were heading for home.

John Michener was flying Clover Red Three and temporarily attached his element to Clover Blue Flight during the general battle. He damaged one Oscar and fired at another without visible results before he noticed one enemy plane fall into Sentani Lake. The B-25s were just making their run and the parafrag bombs were visible as explosions below. While he was rejoining his own flight and heading for Humboldt Bay, Michener observed John Loisel lead Luddington down to strafe luggers in the water near the shore.

Lieutenant Alex Ilnicki was Michener's wingman and also saw Loisel go down after the Luggers. He also managed to look back long enough to see many black columns of smoke rising from the strip at Hollandia. At the base of each dark column were bright flames that seemed to be coming from burning airplanes.

One latecoming P-38 had joined the 432nd formation over Annenberg on the way in to the target. Captain Richard Bong was the leading fighter ace in the Southwest Pacific with twenty-four confirmed kills and was privileged to attach

Another view of a well kept #110. Note the rudimentary demon on the nose.

Jack Fisk and Clarence Rieman sometime during the initial months at Nadzab. Fisk was another hot fighter pilot who was killed in an air accident after the war.

himself to any P-38 unit in action. During the scrap over Sentani Lake he made repeated passes at one elusive Oscar before it caught fire on the underside of the engine and exploded on a hill about fifty yards west of the lake. Joe Forster witnessed the crash of Bong's 25th victory.

With Bong's victory on its roster the 432nd Squadron had an even twelve kills for the mission. The other two squadrons added three more for a total of fifteen Oscars and Tonys for the day.

The victory was not without cost, however. The last flight of Clover Squadron moved in behind Clover Green during the fighting and Cliff Mann noticed two Tonys closing quickly behind his wingman, Joe Barton, a new pilot from Kentucky. Barton dived away to shake off his pursuers and before Mann could do anything he simply disappeared into a cloud below, followed by the enemy.

Climax at Hollandia

The mission to Hollandia was called off for April 4, once again due to the weather. On The next day another force of B-25s was escorted to Hollandia by the Satan's Angels and the wreckage that was formerly a sizable Japanese air base was churned up like so much shoveling through a junkyard.

Air opposition on this mission was reported to be nonexistent. It would appear that the Japanese were in the process of first abandoning the defense of Hollandia and then forsaking the base itself. From the end of April through August 1944 many Japanese would retreat to bases farther along the Dutch New Guinea coast. A good number of these troops would die on the formidable foot trek. Some would ironically escape crashes or bombs at Hollandia only to be lost during casual

strafing along the route to Sarmi or other destinations farther north. Some would succumb to the rigors of the march itself (Sarmi was about 125 miles north along the coast from Hollandia) and others would suffer the indignity of capture.

It was these captured Japanese airmen (primarily from the nearly annihilated 77th Sentai) who first attested to the effectiveness of P-38 formation attacks. The Japanese felt that a P-38 all alone was vulnerable, but that the formation that stayed together was nearly unbeatable. Thus, the principle 475th battle doctrine was vindicated in the eyes of the enemy.

Black Sunday

The last mission to Hollandia on which Japanese interceptors were encountered in any numbers was April 12, 1944. On that day the 80th Fighter Squadron claimed nine Japanese fighters, three of which were eventually credited to Dick Bong to give him a record-breaking victory total of twenty-eight.

Fisk's #191 during the second quarter of 1944

Lent, McGuire, MacDonald and Loisel in mid-1944.

Japanese records admit the loss of one Tony pilot and two others who flew Oscars. It is possible that as many as six other Japanese either bailed out or survived crash landings to give the P-38s a reasonably close margin of verification.

On April 16, 1944 Easter Sunday the final mission to Hollandia before the invasion of the base was scheduled. Forty-eight P-38s of the 475th Group were readied for the main assault that would begin around eleven o'clock in the morning. The 432nd Squadron would escort A-20s, the 433rd would cover the B-25s and, presumably, the 431st would stay with the B-24s.

Four Clover P-38s would start out a bit earlier at 10:30 to escort some Liberator photo-reconnaissance bombers (F-7) that would photograph the bombing results. The F-7s roamed over the target area from 1:20 to 2:00 in the afternoon taking pictures of what was regarded as a highly successful mission. However, Lieutenant Bob Hubner, one of the Clover Flight escort, experienced some sort of mechanical trouble and was forced to abort. Lieutenant Jack Luddington began escorting him back home at about one-thirty.

At the beginning of the day Clarence Rieman thought that the 433rd would not even get off the ground because of the overcast. The day's mission for the squadron was called off and Rieman settled down with his tentmates, Lou Longman, Austin Neely and Bob Tomberg to work on screening in their tent.

By midmorning the sun was shining again and the mission was on. Rieman was a little uneasy about the clouds that did remain, but he was glad to be off on the mission rather than sitting around on the ground waiting for the weather to clear.

Dick Kimball was leading the squadron as Possum Red One. Besides Rieman, Longman, Neely and Tomberg behind him there was Cal Wire leading a flight. Bob Tomberg was leading Possum Blue Flight which included Lieutenant Carroll "Andy" Anderson, who had been with the 433rd since January, and Lieutenants Pierre Schoener and Joe Price.

Fifty-eight B-24s, forty-six B-25s and 118 A-20s were met by the P-38s at about ten thousand feet on the way to the target, south of Wewak. Two of the Possum P-38s were forced to abort with fuel problems and only fourteen of their P-38s were reported weaving over the B-25s as they closed on Humboldt Bay. The B-24s went in first and did a first-class job of damaging the Japanese defenses. There were bright yellow flames coming from several places on the dark green of the jungle and dirty smudges of smoke were staining the azure blue skies.

Possum Blue Flight stayed with the B-25s until the bombs were dropped and the order was given to turn around and go home. This was the one hundredth mission for Possum Blue leader Bob Tomberg and he had a bad feeling about it. Just before the takeoff he mentioned his uncertainty to his crewchief who offered to declare the P-38 non-operational and have Tomberg scrubbed from the mission. Tomberg just smiled sheepishly at his own qualm and answered simply that he would fly.

There were no fighters in the sky and the few puffs of antiaircraft fire coming from batteries that escaped the B-24 attack were not worrisome. The most troubling thing about the orderly retirement from the target was the cloud cover that now closed in behind the Americans. Even that could be disregarded as long as the weather stayed behind them.

With a little extra fuel remaining because of the lack of Japanese air opposition the P-38s were free to frolic a bit on the way home. Dick Kimball led Possum Squadron into a rat race in and around the bombers. Waist and tail gunners on the bombers would wave and cheer as the P-38s darted in and out, flaunting their agility to the admiring bomber crews.

This aerial playfulness had its disadvantage in the extra consumption of fuel. One by one the P-38s would use up the fuel in their external tanks and their engines would windmill for a short time until a gray puff of smoke from the superchargers announced they were running on what was left in the main tanks.

For the two days preceding the April 16 mission the weather services had recommended cancellation of operations because of the threat of severe conditions developing. Now, while the P-38s were frolicking in a remaining pocket of calm weather, a massive front ranging from the Admiralty Islands in the north to the Gulf of Papua, blocking the path home as far north as Saidor.

Calvin Wire had already aborted the mission with his wingman, a new pilot named Mort Ryerson, because of fuel problems. Since he was sweating out his fuel anyway Wire was even more uneasy about the solid overcast above his altitude of 12,000 to 13,000 feet, and the lower broken clouds below. When the two P-38s reached a point south of Wewak and approaching the Owen Stanleys the front became a solid wall of clouds from the ground to as high as the pilots could see.

There was no immediate panic and the old New Guinea veteran Cal Wire led his wingman up to 20,000 feet to avoid the higher peaks of the mountain range. The ominous mass of gray stretched many thousands of feet higher and Wire decided to lead Ryerson around to the north. There was seemingly no way around the great cloudy barrier and the two P-38s chased back and forth looking for any opening.

The fourteen remaining P-38s of the 433rd were nearing the front at about this time. Andy Anderson could hear Wire and Ryerson over the radio as they assessed their situation. He could also catch some conversation between B-24 crews ahead that tipped him off to the "socked-in weather over the Markham Valley." The way to Nadzab was blocked, compelling Anderson to look for an alternate place to land.

This perilous situation was already breaking up the 433rd formation, which, among other squadrons, started to divide into individual flights looking for the safest and fastest place to set down. The airwaves were filled with calls to Gusap for directional homing.

Meanwhile, Wire had noticed a B-24 off to his left and called Ryerson to follow him while he latched onto the big bomber's wing. With a navigator and radioman aboard, the B-24 had a much better chance of making it home than two single-seat fighters. The B-24 rocked its wings in friendly acknowledgement when the two P-38s slipped in on its right wing. For about fifteen minutes this little formation of B-24 and two P-38s cruised along happily. Then, for whatever reason, the B-24 made a sharp left turn and left the P-38s behind. Wire called to Ryerson and ordered him to climb and turn 180 degree in order to exit the way they came in.

It took another ten minutes for Wire to get his bearings and settle down on a homeward course. A full squadron of B-25s then happened along and the two P-38 pilots were overjoyed to once again tuck in on the wing of the trailing bomber. Within a few minutes these bombers flew off in every direction and the P-38s were obliged to repeat the direction reversal that would take them out of the clouds.

Wire was a bit desperate at this point and called Ryerson to suggest that if they flew along the coast this sort of weather would leave about a fifty foot gap above the shoreline, allowing them to scoot toward a friendly landing place. Ryerson agreed to try and they went down to find the shoreline.

They were happy to find that Wire was right when about a forty foot gap appeared over the water near the edge of the churning surf. It was raining hard, so hard that Wire and Ryerson had to use their side windows just to maintain some sort of formation and to avoid the angry froth of the waves a few feet below.

Aircraft #187 of the 433rd coming to grief during a forced landing. Nadzab and Hollandia were both famous for many minor aircraft accidents.

With little warning the coastline that had been consistently on the right suddenly turned out to sea and blocked the path of the P-38s. Wire pulled back on the control yoke, but was too late to avoid a palm tree that nicked his right prop. With some effort he managed to steady his fighter and got out over the surfline again. When his right engine began a heavy shaking and vibrating Wire realized that he had not got off entirely without penalty. He feathered the right propeller and shut the engine down. At about the same time, Ryerson called that he had had enough of this and, since he had enough gas and ample instrument training, would like to try climbing out of the mess. Wire acknowledged and wished him luck.

Wire continued his lonely flight down the coast on single engine. He was "getting awfully tense and . . . downright scared." The only idea he had of his position was that he was flying down the coast approaching Lae.

Suddenly he was relieved to find a small landing field with a number of aircraft on it. With understandable eagerness he turned directly toward the field with a bit too much airspeed and overshot it. He turned back toward the water and, keeping the field just in sight, turned south for a longer approach. Unfortunately, the clouds decided not to cooperate any longer and started descending to water level, forcing Wire closer and closer to the madly churning waves. One of the white-frothed gray mounds of seawater seemed to break higher than the P 38 was flying and before Wire could get its nose above it the Lockheed plowed through the top of the wave.

Bong's P-38 after his 27th victory. He scored his 25th with the 432nd over Hollandia.

SSgt Heap's radio shack. Sometimes the skip waves would get signals from Los Angeles or San Fransisco. A loudspeaker hookup would bring Glenn Miller or Tommy Dorsey to the entire camp. (Heap via Maxwell)

Immediately the fighter went out of control, the nose bouncing up and down and finally heading for the ocean only a few feet below. The wheel came all the way back into Wire's lap; he cut the mixture control and jettisoned the canopy, at the some time bracing for the inevitable impact.

When he came to, it felt as though he had needles piercing his eardrums. He was unnerved to realize that the P-38 was already underwater and he quickly undid his seat belt. There was enough flotation in the parachute and in his flying clothes to raise him the estimated twenty-five feet to the surface after he had kicked himself free of the sinking P-38.

Understandably in a slight panic when he got to the boiling turbulence of the raging waves, he tried to inflate his life raft before he remembered in disgust that he simply had to pull the CO_2 tabs on his Mae West. Gratefully he laid back for a rest after the reassuring whoosh of the cartridges releasing the gas into the life vest told him that he could stop struggling for the moment.

With his strength renewed he was able to inflate and finally get into the raft. He measured the distance to shore at about one half mile. A dugout was coming from the direction

of shore and he spent an anxious few minutes in the fear that the two men in the native boat were Japanese.

Fortunately, the men were American and Wire's .45 Automatic wouldn't fire, for he tried to shoot what he took to be enemy troops. The GIs took their battered comrade to the first-aid station across the field (they had to wait for two B-24s and two B-25s to crashland on the strip before Wire could be treated for a badly broken nose, several serious facial cuts and damage done by his front teeth being driven through his lip.)

Wire learned that he had made it to an L-4 Piper Grasshopper field at Yamai Point, a few miles northwest of Saidor. The next day he was put on an LST to Saldor and was eventually evacuated to a hospital in Sydney. Cal "Bud" Wire was still full of fight, however, and caught up to the 433rd at Biak in August.

The struggle to get down from the April 16 mission was not by any means over and the rest of Possum Squadron was facing a severe problem. Andy Anderson had turned on his artificial horizon and gyro compass, and checked the fuel gauges before leaning out the mixture for maximum endurance.

This page and opposite: Various views, probably of the Markham valley. (Heap via Maxwell)

The weather ahead was appalling in its seeming coverage from the jungle floor to as high as the eye could see. Anderson was watching Bob Tomberg's P-38 with some consternation. Tomberg seemed to be losing his line of flight and Anderson tried to warn him by radio.

"Hello Tomberg. This is Anderson. You are in a spiral to the left. Pick up your left wing."

Tomberg did not answer. His radio was out and so was his artificial horizon. He had only Needle-Ball and Airspeed Indicator to give him a clue to his flight attitude. Possum Blue Flight had entered the weather at 12,000 feet southeast of Madang and Tomberg had first tried to get under the front then climbed to 25,000 feet with no better luck.

He finally settled down to 15,000 feet where the murk seemed even denser and there was an ominous green tinge to the clouds. The fear of crashing into some mountaintop was gripping all members of the flight, especially since Tomberg could not be reached by radio while his P-38 tended to spiral off in one direction or another.

When Possum Blue entered a clear canyon of air that was completely enveloped by clouds Pierre Schoener waved goodbye to his comrades and broke formation to try and make it on his own. Anderson and Price stuck with Tomberg who just looked back at them and shrugged when he entered the weather again.

Finally, Tomberg decided that he was not going to get back unless he bailed out and walked. He was still unaware that the radio was out when he called to Anderson to take over. Destroying the radio equipment, rolling in some trim and stuffing his maps inside his flight suit, he opened the canopy and slipped between the booms, hitting the radio wire before he fell into the gloom below.

Anderson and Price became separated in the nerve-racking race between the green spectres of the mountaintops that appeared with little warning. Eventually, Price made it to Saidor by five o'clock in the afternoon and Anderson made it all the way home to Nadzab with minimum fuel. Ironically, one of his droptanks that he thought had fallen clear was hung up on its underwing shackles and dropped off of its own accord and exploded near the tower on final approach!

Schoener had also made it home as did Jack Daly. Stanley Northrup made it in at Saidor around four o'clock and Mort Ryerson landed there, too, half an hour later. Eventually, Dick Kimball, Al Pomplun, LeRoy Ross, Howard Stiles and

ANNEX 'F'

SECRET

HEADQUARTERS 475th FIGHTER GROUP
APO 713   Unit 1

21 April, 1944.

MEMORANDUM:
)
TO   : Commanding Officer, 431st Fighter Squadron.
                                 432nd Fighter Squadron.
                                 433rd Fighter Squadron.

<u>THE JAPANESE AIR MEDAL</u>

    1. Commanding Officers of Fighter Squadrons of the 475th Fighter Group are hereby authorized and directed to recommend the award to members of their command, of a decoration to be known as the Japanese Air Medal.

    <u>a.</u> The Japanese Air Medal shall be awarded to any member of the 475th Fighter Group who, by his conspicuous poor judgment and un- yielding resistance to the dictates of common sense, has distinguished himself and has made a substantial contribution to the cause of Japan.

    <u>b.</u> Recommendations will be made in writing by the Squadron Commanding Officer to the Commanding Officer of the Group. An original and one copy of the recommendation shall be submitted.

    <u>c.</u> The recommendation shall set forth the benefit derived by the Imperial Japanese Forces in terms of damaged equipment, time lost and general wastage because of the occurrence. The technical deficiencies and mental aberrations which produced the event shall be sufficiently described in the hope that others, saddened and instructed by the calamity, will forever shun the practices which made it possible.

    <u>d.</u> Whenever it appears that a man has become eligible for the award of the Japanese Air Medal, the Squadron Operations Officer and Engineering Officer shall immediately investigate the matter carefully and report their opinion to the Squadron Commanding Officer. If the award has been earned, a proper recommendation will be promptly submitted.

    <u>e.</u> If the Group Commander approves the recommendation, a citation will be issued and the decoration presented as promptly as possible.

SECRET

    <u>f.</u> In the deplorable event that the award is earned more than once, no new medal will be presented, but the awardee, after issuance of the citation shall present his medal for the addition of a suitable shingle to the Hon. Air Medal, (H.A.M.).

    <u>g.</u> The medal shall be worn over the right breast pocket daily from the hours of 1700/K to 2100/K, for the period (one week) prescribed in the citation.

                 By Order of Lt. Colonel MACDONALD:

                                   /s/A. R. FERNANDEZ
                                   A. R. FERNANDEZ,
                                   Major, Air Corps,
                                   Adjutant.

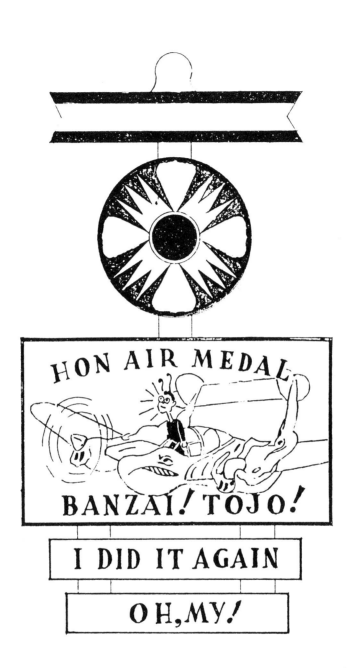

Clarence Rieman all made it back to Nadzab, but Longman, Yarbrough and Neely were lost somewhere in the unyielding New Guinea countryside.

Things were bad for Clover Squadron, too. Captain Loisel managed to keep the squadron together for a time, but the fear of collision separated the flights. Somehow, the sixteen P-38s of the main escort all managed to get down safely, but the two aborting Clover ships of the F-7 escort got into deep trouble.

Loisel got into radio contact with Lieutenant Hubner and found that he was in distress somewhere to the front. It was apparent that Hubner was extremely shaken and in no condition to fly on instruments. He was advised by Loisel to bail out, but seemed much too nervous for any such suggestions.

Art Peregoy also heard Hubner over the radio calling for a direction from "Harvest", one of the homing fields around Saidor. When his voice over the radio faded out Hubner just slipped out of communication with any American source. Both he and Luddington disappeared into the jaws of the weather and were lost forever.

The 431st Squadron got off a little easier than the other 475th units. Lieutenant Milton MacDonald was the one casualty listed for the day, giving the group a bitter loss of six pilots in an otherwise uneventful mission. General Kenney's Fifth Air Force lost a total of thirty-one aircraft and thirty-two aircrew for one of its most stunningly painful days.

Bob Tomberg's Trek

First Lieutenant Robert Tomberg was surprisingly soft-spoken for such a scrappy young pilot. He already had two Japanese aircraft to his credit when he decided to bail out into the murky clouds on April 16, 1944. Whatever lay beneath him now had to be more negotiable than ramming a mountaintop in a P-38 with malfunctioning instruments.

He drifted out of the clouds at about four hundred feet above the trees and tried to aim for any decently open ground. For just a moment he caught sight of a small clearing with a native garden before the treetops came rudely up to meet him.

The parachute caught itself on a limb 35 to 50 feet above the ground. Tomberg's knee was badly wrenched and gashed, forcing him to gingerly make his way down the closely packed tree trunks. Arduously, he descended to the ground and looked around for the clearing and native garden.

At first he tried to go down a steep embankment, but when he tossed a stone through the green foliage and didn't hear it land for about twenty seconds, he decided to try another route. His luck was good and he found a small trail. A sudden grunting noise put him on alert and he found that he was pointing his .45 automatic at a wild pig.

Circumventing the animal on the trail, he was happy to find the native garden with a small fence surrounding it. In the garden was a young native girl who seemed momentarily unaware of his presence. He formulated the appropriate pidgin-english phrase in his mind, but when he spoke the girl was startled and ran.

Rain was still beating down on the exhausted pilot and he found what shelter he could for the night. The next morning he found footprints that he recognized as the peculiar footwear of the Japanese and decided to struggle up into the trees as his best means of protection.

Again his luck held and the first people he saw coming toward his perch were some natives. The leader was wearing a red hat and something told Tomberg that this was a potential rescue. He called to them when they approached the base of his tree and made his way down.

Communication was cautious at first, but the leader with the red hat told Tomberg, "Japan-man, him go Madang." Whatever suspicions existed eventually melted and the natives made friends with the downed American pilot and took him to their village.

Tomberg experienced the local custom of greeting at the village. All the men of the community lined up for a formal handshaking ceremony, the chief first and then the rest of the men down to some teenage boys. After the ceremony was completed, medication was applied to Tomberg's knee that seemed to assuage the pain and he was given some sort of potato soup that tasted good after a day without food.

The next morning Tomberg set off with four native guides two ahead and two behind him for the next village. The handshaking ritual was repeated when he arrived there, once again with only the men even though he caught a glimpse of a woman in the doorway of a hut.

Again he was given food and rested before setting out with native guides in the direction of Saidor. He was careful to take periodic doses of atabrine tablets which aroused the curiosity of his companions. Thinking of it as a goodwill gesture, he gave some of the tablets to the guides.

One of the services that the guides performed for him was to carry him across the rapidly moving streams. A slight accident in one of the shallower streams caused him to tumble into the cool and rapidly rushing water. It was such a refreshing change from the hot and steaming forest that Tomberg demanded to stay in the stream for about fifteen minutes.

After about five days in the New Guinea brush Tomberg reached the coast with his guides. He now experienced the famous six foot tall kunai grass that sometimes limited the progress of downed airmen to a mere few hundred yards a day. While the going from tree grove to kunai grass back to tree grove was difficult the guides were expert at moving through the countryside and Tomberg had a machete for the thickest terrain.

At about this time Tomberg had the worst fright of his journey. While passing through one of the groves he heard a crack and looked up to see one of the limbs of a tree just ahead fall into the brush near the trail. A huge twenty-foot snake that had been waiting for its dinner to pass beneath the limb simply broke the branch with its great weight. Tomberg fired two shots from his .45 into the brush where the snake vanished and held his machete in front of himself from that point on through the rest of his trek.

Not long afterward Tomberg reached a point where he met a representative from Saidor. The guides were given enough of a reward to buy two wives or one pig and Tomberg made it into Saidor where he wangled a position as tail gunner on an A-20 heading back to Finschhafen.

At Finschhafen he met more opposition than he had found on the jungle trail. Base Operations refused to give him a seat on any flight to Nadzab because he had no orders! A friendly Tech Sergeant advised him that a C-47 was preparing to take off and that Tomberg should talk informally to the pilot.

Tomberg took his advice and got a hitch on the C-47 to Nadzab. However, when the battered P-38 pilot finally did arrive he was stranded again for the lack of transport to the other side of the base. He tried to call the 433rd Ops without success, but again managed to hitch a ride with a sympathetic officer who was going in Tomberg's general direction with a supply truck.

It had been seven days since he parachuted into the New Guinea jungle. He was tired and hurting from the rigors of the trek and the injuries he sustained in the jump. Even the small creature comforts of his tent and the 433rd mess at Nadzab seemed delicious at the moment. Somehow, the desperation of his plight inspired him to convince the officer to go out of the way to the 433rd Campsite.

Tomberg got out of the truck when it screeched to a halt in the 433rd area and thanked his benefactor sincerely. After a week of being missing in action, he was pleased to find his tentmate Clarence Rieman sitting inside their canvas home away from home.

Rieman simply looked up with a half-smile and asked, "Where the hell have you been?"

Chapter X

SATAN'S ANGELS OVER WATER MAY-JUNE 1944

"Between April and August 1944, events in the Pacific moved at breathless pace. General MacArthur's Southwest Pacific Forces, mostly American but with Australian Naval and Aviation Engineer increments, made the landings that secured Hollandia in Dutch New Guinea on 22 April, established a new forward base and airdrome, and, in a series of well conceived and smartly executed moves, took Wakde, Biak, Noemfoor and Sansapor."*

So did the eminent U.S. Naval historian, Admiral Samuel Eliot Morison, give credit to General Douglas MacArthur's final drive up the north coast of New Guinea. The landing at Hollandia finally broke the back of Japanese resistance and spelled the end of operations for several venerable air units, including the 33rd and 77th Sentai with their Oscars and the 68th and 78th Sentai with their Tonys.

During the period of the Hollandia raids the 475th was the only V Fighter Command P-38 group. The only other P-38 squadron in New Guinea at the time was the 80th which is listed on some organizational charts as operating attached to the 475th rather than its parent 8th Fighter Group. In the period between February and the end of April 1944, the 9th Fighter Squadron reconverted from P-47s to P-38s and the rest of the 8th Fighter Group received theirs to form a second complete P-38 group.

By the time the Japanese retreated in some disarray from the Hollandia area there were, therefore, seven P-38 squadrons ready for the next assignments of Fifth Air Force. Those assignments would be strategic in attacking oil and airfield targets on the way to the Philippines. The assignments would also be tactical in mopping up the remaining Japanese forces on New Guinea's Vogelkop Peninsula.

The Navy in the Central Pacific and MacArthur in the Southwest Pacific were now converging on the Philippines to form an iron ring around Japan's secondary defense perimeter. The final defense line would be breached in February 1945 with the assault on Iwo Jima and in April with ultimate landings on Okinawa and its satellite island, Ie Shima.

By June 16, 1944 Tom McGuire had Twenty confirmed victories. Dennis Cooper made him part with his infamous service hat for this photo. If the hat wasn't in the cockpit when McGuire was ready to fly, there was the devil to pay. (Cooper)

During May 1944 the 475th's P-38s were advanced to the newly won bases at Hollandia. The stay at Hollandia was supposed to be a short one before the island of Biak, farther north in Geelvink Bay, was taken. As it happened, the fight to take Biak was much stiffer than the token resistance experienced at Hollandia, and the 475th remained at Hollandia until the middle of July.

475th camps at Nadzab were broken up on the first day of May 1944 and equipment was sent to the staging area at Lae, via the Liberty ship, "SS Francis A. Wardwell." The ground echelon moved to Lae by truck and car.

*NOTE: Samuel Eliot Morison, *History of United States Naval Operations in World War II*, Vol. 12, "Leyte."

Remains of Ki-51 Sonia at Hollandia. (Hanks)

The capture of Hollandia by the Allies had been so swift that many Japanese aircraft were simply awaiting maintenance, others were in packing crates that indicated they had just arrived.

The air echelon remained at Nadzab and waited in the skeleton of the camps while the Lae contingent enjoyed swimming at the beach during the day and movies or the Red Cross canteen in the evening. It was the first time that the three squadrons of the 475th were encamped together with a combined mess arrangement in the Lae area.

On May 1, John Loisel was promoted to Major and Art Peregoy was rotated home on May 13; the 432nd Squadron air echelon was ordered to Hollandia two days later ahead of the Lae boat echelon. It appeared that the Nadzab contingent would have to suffer deprivation just a little longer.

As spartan as the Hollandia bases started out to be the first thing many troops noticed was that it was possible to stand ankle deep in mud with dust blowing in your face, missions

Apparently three Japanese infiltrators were unsuccessful in trying to cross the 475th perimeter on May 11, 1944 while the group was preparing to move to Hollandia.

were started as early as May 16. Initial operations were escorts to Biak in preparation for the invasion beginning May 27.

There were some determined attempts to make the Hollandia camps more livable before the boat echelon made its way to the shores of Humboldt Bay. Major Claude Stubbs was a first class Supply Officer and midnight procurer whom Colonel MacDonald had influenced to come into the 475th from another group. He was working on the Hollandia campsite and doing well in spite of the occasional competition from starving Japanese stragglers, two of whom he had to chase away with his .45 automatic on the site at Ebeli Plantation.

Another desperate Japanese straggler ventured into the 475th camp, probably in search of food, and came face to face with one of the air echelon Satan's Angels pilots. The startled pilot seemed less than satanic when he burst out in a reflex, "Who the hell are you? What the hell are you doing here? Get the hell out of here!"

It was frightening enough for the Japanese who turned and ran back into the bush. Whoever the 475th pilot was, he was lucky enough not to find his gun until he and his aroused fellow pilots were calm enough, else they could have shot each other as soon as the Japanese interloper.

The boat echelon did not embark until the first of June and arrived in Humboldt Bay around the sixth of the month. The Governor of Dutch New Guinea had a view of the entire fleet of ships dropping anchor practically at his feet. He estimated the number of ships to be about five hundred and the number of Yanks pouring into his front yard to be infinite.

He had the grandest house in the area, a fact not unnoticed by MacArthur's staff who occupied it after the American invasion was stabilized. It was from there that MacArthur worked out the final details of the Philippines operation.

431st flight including the P-38s of "Pappy" Cline, Tom McGuire and Bill O'Brien. This photo was probably taken around May of 1944 and the 431st still maintains the practice of placing unit numbers on P-38 nose.

The first thing that many of the groundcrewmen noticed about the abandoned Japanese camps was the filthy trash heaps left by a hastily retreating enemy. That and the bits of evidence of on site brothels – an arrangement that no American airman would tolerate – proved to some minds that they were fighting a definitely heathen foe.

Some of the airfield facilities were quite usable. There were in evidence huge stockpiles of equipment and supplies, proving again that the Hollandia operation came as a surprise to an enemy who considered the base quite safe only a few months before.

There were new Nakajima Ha-25 radial engines still in crates, stacked in long rows. Other rows had spare parts for aircraft and trucks, while some vehicles and aircraft stood completely assembled. Some of the aircraft were still packed in cosmoline, which fact indicated that they had arrived assembled and were never used.

First Combat From Hollandia

Warren Lewis led sixteen Possum Squadron P-38s on an F-7 Reconnaissance mission to Kamiri Airdrome on Noemfoor

P-38 #136 abreast of an ex-enemy at Hollandia.

Major Oliver McAfee of 475th Headquarters usually flew #102 behind him. (475th history)

Island when he ran into the first aerial combat from Hollandia on May 16. The sixteen P-38s were divided into two groups of eight, one at 15,000 feet and the other at 18,000.

The P-38s rendezvoused with the photo-bombers at one fifteen in the afternoon and proceeded to the target. An hour later three Oscars were sighted down below at 5.000 feet, just as they were taking off from Kamiri Airdrome. Lewis ordered tanks dropped and went down to the attack with the lower two flights of P-38s, leaving the other eight as high cover.

Lieutenant John Purdy was leading the second element of Lewis's flight and followed him down willingly. Purdy had been with the squadron since December and was known as an eager beaver who could be depended upon to stick with his leader. He now watched while Lewis and his wingman, Howard Stiles, one of the lucky ones who made it back to Nadzab during "Black Sunday", made a pass at two of the Oscars.

Lewis made a head on pass at the lead Oscar. Purdy saw a two second burst come from the guns of his flight leader and explode the Oscar which then fell into the water below. Stiles fired at his and it, too, was shot down. All three pilots then started a chase of the third Oscar that ended up in deadly game of tag.

Purdy heard Lewis call that he was out of ammunition after making a number of frustrated passes at the agile Japanese fighter. With his chance now at hand, Purdy came down from above on the Oscar which was at the moment right down

Headquarters P-38 #101 was usually flown by Meryl Smith. Note the 432nd clover leaf on the nose.

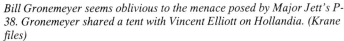

Bill Gronemeyer seems oblivious to the menace posed by Major Jett's P-38. Gronemeyer shared a tent with Vincent Elliott on Hollandia. (Krane files)

Fearsome devil's head motif on the nose of a 431st P-38.

on the water. The P-38 drove in from behind and flamed the hapless Oscar which was seen to splash into the water southwest of Noemfoor.

Some of the B-24s dropped bombs in the combination recon/bombing mission and the Americans could see smoke or dust rising from hits on the airdrome. The P-38s broke escort not long after the combat ended. They could head home with the satisfied feeling that their gun-camera films detailed the destruction of the three dark green Oscars.

McGuire Back In Action

Sometime during the move from Nadzab to Hollandia Tom McGuire took command of the 431st Squadron. He was also promoted to Major at the same time and apparently felt the pressure to accomplish even greater feats of daring in combat. Beginning from the Hollandia days, the notes of displeasure increase in regard to his attitude. The point of fact may actually involve the increased responsibility he faced as commander of a Satan's Angel squadron. Whatever else he may have been, McGuire was dedicated to the record of the group and ruled to the letter of the law.

On May 17 he was leading sixteen P-38s on an escort of B-24s to Biak when Hades Squadron ran into four Oscars. He

managed several deflection shots and set one Japanese aircraft on fire. Oil streaming back from the Oscar slapped onto McGuire's windscreen, but he was able to watch his victim crash into the sea. Another Oscar that McGuire was chasing was shot down by Captain Bill O'Brien and John Tilley got a third Oscar for his second victory.

Two days later McGuire scored his eighteenth official kill when he led another escort of B-24s to the island of Manokwari. Twelve Hades P-38s arrived over the target, once again at the advantageous altitude of 18,000 feet. Japanese flak was unusually heavy and accurate with black bursts appearing like a carpet between 14,000 feet and the escort altitude.

McGuire was leading Hades Red Flight and the bombers were approaching the target area when two flights of Ki-44 Tojo radial-engine fighters came diving down from the right. They crossed the line of flight of the Hades P-38s and McGuire immediately ordered tanks dropped.

At first McGuire was a little surprised by the appearance of the enemy fighters, but quickly recovered in time to slip in on the tail of the lead element. His target was a dark camouflaged Tojo that showed signs of hits from his first burst. The Japanese zoomed up into a stall and exploded after a second burst. The force of the explosion blew the pilot clear of the cockpit and he parachuted into the water.

Jim Moering inherited Frank Lent's #134 during the middle of 1944. (via Bob Rocker)

Calls were coming back from the bombers that the other flight of Tojos was dangerously close and McGuire turned back to his escort duties. He learned later that a formation of 9th Fighter Squadron P-38s claimed all three remaining Tojos of the flight that he had attacked. Incidentally, Warren Lewis claimed his seventh victory, a Mitsubishi "Pete" recon bi-plane, on an A-20 escort to Noemfoor earlier in the day.

Life At Hollandia

Some of the pleasures at Hollandia overshadowed the slightly disquieting sounds of gunfire coming from the hills where mopping up operations against Japanese pockets of resistance were being conducted. The 475th camp was shaded by the tall coconut trees that filtered the sunlight but allowed fresh breezes to waft gently through. A cold, clear stream flowed next to the camp – unlike the torrent of the Erap River – and cold showers waited to relieve the Satan's Angels from the heat of the day.

Since it became evident that Hollandia would be a longer stay than anticipated after the Biak operation became bogged down at the end of May, many of the conveniences of rear area life began reaching the forward bases. However, the complaint of the enlisted men in the squadrons was that some of the little considerations were not reaching the 475th. Very little beer was issued to the other units in the area while the 475th got almost none at all. Promised issues of cigarettes were fulfilled by the inferior and climate-damaged captured Japanese stocks.

It was obvious that someone else was enjoying the intended rations, perhaps purchased on the black market. One

item that did manage to get through was the almost unmeltable G.I. issue chocolate bar. Unwrapping the brown log that stayed solid in even 100 degree heat was a chore in itself, but most troops who persevered relished the rancid-tasting treats.

Accidents and Attrition

From the terrible cost of the Bloody Sunday mission in April to the end of May there had been virtually no casualties in the air for the 475th Group. To be more accurate, more damage was done by Satan's Angels pilots than by the enemy to the group's P-38s.

With more pilots being rotated home throughout the first months of 1944 there were a large number of new people in the squadrons who needed training in theater operations. Accidents involving landing gear leaving wheels up on landing or damaging struts on hard handing, for example were common throughout the period up to the Hollandia move.

Colonel MacDonald used a clever ploy to encourage speedy development of P-38 operational proficiency. He issued a "Japanese Air Medal" to anyone damaging a P-38 through less than standard procedure, and the embarrassing award seemed to work (see pages 144-145). During May there were only three P-38s transferred to service squadrons, one of those being the aircraft that was damaged by flak when it was flown on the May 19 mission by Captain Campbell Wilson.

June did see some bad accidents, however, that could have had more serious results than they actually did. On June 16, 433rd Squadron pilot Lieutenant John Knecht ran out of gas during an emergency landing at Hollandia and bailed out at five hundred feet over a crocodile infested part of Lake

One of the spoils of war at Hollandia; a groundcrewman pedals past #170 on what appears to be a liberated American bike that was originally liberated by the Japanese. (Anderson)

Warren Lewis got credit for a Pete recon biplane on the same day that McGuire got his eighteenth victory, May 19, 1944. (Lewis)

Sentani. His fellow pilots rushed to his rescue only to find that he had landed on the only spit of dry ground in the area and was having cocoa with a refugee Javanese lady and her husband when they located him!

Captain Billy Gresham had a more sobering experience on June 21 during a local test hop. He took a P-38 up in the afternoon and made a crashlanding on Hollandia Strip which left him with facial lacerations and some broken teeth.

About the same time, Fred Champlin also had a bad flying accident. He was just taking off for a cross-country flight to the island of Wakde when an engine cut out and he was forced to make a hard landing. Fortunately, both Gresham and Champlin got off with relatively light injuries.

Combat operations were beginning to take their toll again during June. One the June 16 mission to Jefman Island several group P-38s were damaged. Oliver McAfee from Headquarters and Robert Hadley of the 432nd Squadron brought home P-38s damaged by 20 mm fire from enemy aircraft, while George Veit of the 431st Squadron came home in a fighter that

was put out of service by his own released drop tanks. Howard Stiles, the 433rd pilot who made it home from the Black Sunday ordeal, was shot down on the same mission.

Biscuit Bombers Again

Aside from the usual patrols between Hollandia and Sarmi, the Satan's Angels returned to the mundane job of escorting the "Biscuit Bomber" transport aircraft to Wakde in May. Paul Lucas was on one of the Biscuit Bomber missions when he sighted dozens of Japanese soldiers crossing a small stream at Karboes, on the coast about midway between Hollandia and Wakde. Lucas was returning early from the mission when he saw the enemy troops and made two strafing passes before continuing home on the deck along the coast.

A young and energetic John Purdy got his first victory over Noemfoor on May 16, 1944. (Purdy)

Resuming Aerial Combat

With many of their combat units wiped out after the Hollandia operation the Japanese had to reorganize their defense lines to counter both the Southwest and Central Pacific Allied thrusts which were converging above New Guinea. The Imperial Japanese Navy would be in contact with V Fighter Command P-38s on a more active basis during and after June 1944.

The Japanese 23rd Air Flotilla consisted of about twenty Zeros and a dozen land based bombers around the Vogelkop area on June 1, 1944. They were soon reinforced by seventy Zeros and sixteen D4Y "Judy" divebombers. Japanese Army air units were badly disorganized at the time and made only sporadic appearances.

The first contact that Satan's Angels P-38s had with Japanese aircraft since the battle of May 19 was during an escort of A-20s to Manokwari on June 4. Actually, the contact was so brief and simple that most of the fifteen P-38s in the Hades Squadron formation didn't even drop their tanks.

Tom McGuire was leading Hades Squadron on the uneventful bomb run and retirement from the target area when the A-20s were left on their own to safely return home. Paul Morriss was leading the third flight when about five minutes later he saw a lone Oscar flitting along the deck on the east coast of Mois Neom Island, between Manokwari and Japen Island.

He called out the enemy plane to McGuire who once again was frustrated when he couldn't locate the aircraft flying below. Morriss didn't wait for McGuire, but decided to start down with Lieutenant Bob Crosswait, his element leader who had come up on Morriss's wing to cover him.

For a short time the Oscar disappeared under a cloud before Morriss continued on and rolled in to get close on its tail. One short burst and both P-38 pilots watched the Oscar roll off to the left and crash into the sea.

"Pete" Madison nearly paid all his dues on the Manokwari mission of June 6, 1944. He reported much later, "There I wuz, fifty feet off the deck, upside down. My engine is hanging out and blood, all mine, is all over the place. Am I downhearted? Am I in a state of despair? Am I ready to give up? YES!"

Air Action Over Waigeo

The Japanese were taking some heart over the prolonged struggle at Biak. Some 2,500 troops of the 2nd Amphibious Brigade were loaded aboard transports at Mindanao and rushed toward Biak by the first days of June. However, sightings by American submarine and B-24 observers and the mistaken notion that a strong Naval force was guarding the Biak battleground caused the landing operation to be called off.

The unfortunate Japanese ships were not able to immediately turn around since the American forces were alerted and chased the convoy around Waigeo Island throughout the fourth and fifth of June. Certainly the alarm caused in Allied Headquarters was justified since the Japanese could very well have landed the troops and played havoc with the strong escort of warships. There was nothing more than a single destroyer at Biak to oppose the Japanese landing.

Frantic alerts kept Fifth Air Force B-25s shuttling to Wakde Island in the hope that the Japanese convoy could be intercepted and destroyed before it reached Biak. On June 7, Colonel MacDonald led seventeen P-38s of the 432nd Squadron on what was supposed to be an escort of B-25s on the shipping strike to Waigeo. The strike was canceled and the Clover P-38s landed at Wakde where Elliot Summer crashlanded successfully when his nosewheel collapsed.

Early on June 8 the Japanese convoy made a bid to reach Biak and committed itself to an open course long enough for ten B-25s covered by eighteen P-38s of Clover Squadron to attack around midday. Once again Colonel MacDonald led the P-38s off at 7:35 in the morning, landing at Wakde about an hour later. One of the P-38s had turned back before reaching Biak, leaving seventeen to continue the escort from Wakde at 10:30.

MacDonald led his P-38s (which also included several from Possum Squadron) through poor weather to the rendezvous point. About sixty miles northwest of Manokwari the white wakes of seven ships of the Japanese convoy were sighted on the gray/blue water; the weather over the target area was absolutely clear and the attack was unimpeded.

While the B-25s turned off to the left to make their runs on the ships, the P-38s climbed to the right from the same level as the bombers to draw off antiaircraft fire. MacDonald sighted two Zeros above him and started climbing to meet them. Other P-38 pilots started calling out six to eight Zeros and Oscars vectoring in like sharks to get at the B-25s which were even now on their bomb runs through flak thrown up by the Japanese ships below.

Biding his time, MacDonald decided to take on another

Perry Dahl had a lucky break when he shot down his sixth victory and had another dicy landing on June 8, 1944 to escape unscathed again.

flight of four Zeros to his left. The Japanese formation leader hesitated long enough on making a head on pass to allow MacDonald a 30 degree deflection burst before the two planes passed each other. Both MacDonald and Zach Dean saw the Zero burst into flame and crash.

Apparently the Zeros were trying to draw the P-38s into the antiaircraft fire above the destroyers because MacDonald's kill splashed into the water right in the Japanese convoy's midst. The disheartened Japanese fighters broke and ran with MacDonald hot on the tail of what he took to be an Oscar. With obvious desperation the enemy pilot led the P-38 pursuer down to cloud cover at 3,000 feet in an effort to lose him.

MacDonald pressed home his attack anyway and fired a short burst that started the Oscar smoking; one cannon shot had badly damaged one of the wings which shed large bits of metal. The P-38 followed the Japanese fighter in and out of the clouds until MacDonald saw a chance for a clear shot. The Oscar practically turned on its heel and went back into the clouds with MacDonald's P-38 following all the time.

At last, MacDonald followed his quarry into the mist of the clouds, firing all the time and noticing a bright orange

A defiant-looking Bill Gronemeyer by "Little Grace" sometime before his last aerial victory on June 16, 1944. (#121, serial 42-104021)

explosion just as the Oscar vanished into the darkness. A frustrated P-38 ace came out the other side of the clouds and mentally put in a claim for a probable victory when he could no longer find his enemy target.

Perry Dahl was flying element leader in Clover Blue Fight with Lieutenant LeBaron as his wingman when enemy fighters were sighted to the south. Dahl received permission from Clover Blue Leader to attack and rolled over to go after another Oscar.

The Oscar made for the cover of the clouds, but Dahl got in some high deflection shots that starting hitting their target just before the sanctuary was reached. Major Loisel watched from above and saw the Oscar begin to smoke heavily before it disappeared into the weather.

Four more Japanese fighters came out of the same cloudbank, obviously diving to attack the B-25s below. Dahl called Clover Blue Leader who had been trying to follow him, but was out of contact for the moment. With LeBaron still faithfully on his wing, Dahl managed to chase the four Japanese back into the clouds.

Another stray Oscar drew Dahl's attention and the chase began again. Just off the coast of Waigeo Island Dahl caught up with the seemingly unaware enemy and fired a burst from

Gronemeyer's "Little Grace II" with five victory marks after the June 16, 1944 mission. (Krane files)

"Little Grace II" taking off from Owi Island on June 19, 1944 after receiving minor repairs. (U.S. Army)

dead astern. Both Dahl and LeBaron watched the Oscar crash in flames.

Deciding to go home with LeBaron, Dahl flew right over the convoy and saw flak bursting around him. Also below he could see four ships apparently on fire and sinking. He didn't know it then but the ten B-25s had actually sunk one ship and damaged three others, but lost three of their own number with every bomber that survived coming home badly damaged in the courageous assault on the destroyers.

The day wasn't over for Dahl, however, and after this mission he acquired the additional nickname of "Lucky" Dahl. Flying back through and around the same bad weather that the squadron had endured on the way to the convoy, When he was ready to land he found that his wheels would not lower with the normal hydraulic system. His hydraulic pump was also out of commission so he pumped the wheels down by hand and the light indicator told him that the wheels were down and locked.

When he made his landing and the wheels touched the ground the left gear collapsed and caused the P-38 to swerve off the runway where it crashed into the lines of trees. What should have been a horrible accident became nothing more than an incident when Dahl simply walked away without injury.

Two days before, one of the new 431st pilots, Lieutenant Harold "Pete" Madison had a similar bit of luck. He was on an A-20 anti-shipping escort to Manokwari when he decided to strafe a Japanese Fox-Tare (large transport). The enthusiastic Madison got what is known as "target fixation" and concentrated too hard on his run, smacking into one of the ships large

booms and leaving pieces of his P-38 scattered on the enemy ship deck.

He found himself fifty feet off the water, flying upside down with his own blood spattered on the inside of the canopy. Somehow his youthful spirit managed to right the P-38 and he was escorted home by Lieutenant Hal Gray. He got his plane down safely, but it had to be transferred to a service squadron.

For all the courage of the B-25 crews and the skill of the P-38 pilots, the Japanese convoy of June 8, 1944 did not turn and disappear until the threat of an American destroyer group rushed in to deal with the issue made the enemy change his mind. The P-38s claimed three confirmed (the third was an Oscar shot down by Lieutenant Clifford Mann about thirty miles west of Cape Waios) and two others probably destroyed.

The Odyssey of Bill Hasty Part One

One mission flown by the 433rd Squadron on June 8 did not have combat as its objective, but sought to rescue one of its own members who went down three days before. Eight Possum Squadron P-38s ranged out over the Babo area of the Vogelkp Peninsula to find Lieutenant Bill Hasty who had bailed out during and A-20 escort on June 5. Babo Airfield as well as the entire area south of Babo where Hasty was seen to land was scoured; all streams, rivers and native villages were buzzed with no luck.

Hasty had set out with the 433rd Squadron to provide medium cover at 14,000 feet while other P-38s of the 431st Squadron flew low cover approaching the target at about ten

432nd Squadron P-38 #158. J.W. Stein was the usual crewchief.

A.J. Pinkerton was the crewchief of #159.

o'clock in the morning. The flak was especially intense, "bursting all over the place with those familiar black puffs," in Hasty's words.

Sometime around twenty minutes after ten there was a brief shudder in Hasty's brand new P-38J-15 and he lost power in the left engine. Other pilots in the formation noticed smoke coming from both engines just before the left one burst into flames. Hasty knew that the situation was hopeless when the fire began to spread in the wing area and he decided to bail out.

The other pilots in the 433rd formation saw Hasty get out of his cockpit and his parachute open when he was clear of the burning P-38. They noted that he landed safely in a mangrove swamp six miles southeast by south of Babo itself.

On the ground Hasty was less sure of things. He knew that he was only about twenty miles south of the Babo Airdrome, much too close to the inevitable Japanese patrols that were certainly in the area. He had even spotted some Japanese soldiers and natives along the river on his way down.

He immediately hid his parachute to avoid getting one of the patrols on his trail while he made his way to the river in the hope of crossing it after dark and making his way to the coast for rescue. To hide himself during the day he burrowed into the jungle mud and underbrush, waiting nervously for the dark.

His plans went badly awry, however, a few hours later when a bayonet at the end of a rifle pressed against his stomach and a pistol at either side of him pronounced the end of his escape plans. Japanese soldiers roused him to his feet and began to beat him about the head and face after his hands were tied behind his back.

One malevolent soldier who seemed to be the NCO in charge smashed the butt of his rifle into Hasty's face and knocked out a tooth. Abjectly helpless, the American could do nothing about the beating until the Japanese vented their rage and blindfolded him. They led him to a small boat on a dock area of the river, and he guessed that he was on his way to Babo.

Group Logo, three squadron insignia.

Colonel George Prentice after he took command of the 475th FG. 431st Squadron P-38 behind. (Cooper)

General Freddie Smith, Colonels Kearby and MacDonald at an awards ceremony. (Cooper via Larry Hickey)

Above: Movie stars Una Merkel (standing) and Phyllis Brooks (front, squatting) at Dobodura with the 475th during a USO tour. Note the dark prop spinner in this and the preceding photo. At some point in the operational life of the 431st Squadron a dark blue spinner was used on a number of P-38s. Apparently the practice was dropped before the squadron left Dobo. (Cooper)

Right: Gary Cooper in USMC cap. He went for his first P-38 ride with the 475th at Dobodura. (Cooper via Elickey)

Below: 431st FS commander Major Franklin "Nick" Nichols. (Cooper via Hickey)

Harry Woods, the first pilot to become an ace in the 475th. (Cooper via Hickey)

Bill Haning, 431st Operations Officer

Major Nichols sports a victorious pose over an ex-59th Sentai Ki-43 Oscar at Hollandia. (Cooper via Hickey)

McGuire with PUDGY III at the end of May – beginning of June 1944. (Cooper via Hickey)

Lt. Wood D. Clodfelter between February 3 and February 14, 1944. P-38 #124. (Lee D. Richardson via Hickey)

Final PUTT PUTT MARU sometime in mid-1945. (Via Hickey)

Nelson, Col. MacDonald and Col. Loisel by Putt Putt Maru in January 1945.(Krane)

Joe Forster sitting in his P-38 cockpit sometime during the Philippine campaign, 1944-1945. (Forster)

McGuire memorial at McGuire AFB. Remarkably accurate except for parentheses and vertical command stripes. (Kulikowski)

475th Fighter Group Aircraft Profiles

1. 431st FS. P-38H-5 42-66764
Pilot: Bellows Crew Chief: Unknown
This was undoubtedly the first P-38 assigned to Don Bellows after the shared #116 with Sgt Allan as crewchief. Bellows probably got this P-38 later in 1943 and used it at least until the beginning of 1944.

2. 431st FS. P-38H-5 42-66827
Pilot: Kirby Crew Chief: Unknown
Kirby remembers that this P-38 had a solid red nose and prop spinners. It is likely that at least for a brief period saw the use of dark blue front and red rear spinners. Photo snows a faint image of a native head by the aircraft number and name and Kirby confirms that the image was there.

3. 432nd FS. P-38H-1 42-66573
Pilots: Harris/Wilson Crew Chief: Unknown
Both Fred Harris and Campbell Wilson were known to have flown #142 and Harris is reported to have scored at least one air victory in 42-66573. This P-38 seems to have survived at least throughout the Rabaul operations.

4. 432nd FS. P-38H s/n Unknown
Pilot: Peregoy Crew Chief: A. J. Pinkerton
Art Peregoy probably used this P-38 when he scored his victory of November 2, 1943. The P-38 probably survived until the first quarter of 1944 during which time Peregoy received a P-38J.

5. 433rd FS. P-38H s/n Unknown
Pilot: C. Grice Crew Chief: McConnell
Several 39th Fighter Squadron P-38s were reassigned to the 475th in the period around the end of the Rabaul operations. This P-38H was obviously once in the 39th by virtue of its blue spinners and white shark teeth motif. Charles Grice evidently was assigned to this former example of his old squadron.

6. 433rd FS. P-38H-5 42-66752
Pilot: D. Roberts Crew Chief: T. Hanks
Roberts used this P-38 during the short time that he was commander of the 433rd. He was lost in another P-38 on November 9, 1943 and his usual mount apparently lasted at least for a short time in the 433rd.

7. 431st FS. P-38J-10 42-67597
Pilot: Champlin Crew Chief: Unknown
Fred Champlin used this P-38 for only a short time at the beginning of 1944. He was so dissatisfied with it that he quickly took another for his own and also adopted the number 112. 431st legend has it that no other P-38 was ever assigned the number 113.

8. 431st FS. P-38L-5 44-25638
Pilot: DuMontier Crew Chief: E. R. Lowry
Louis DuMontier got this P-38L around April 1945 and used it for the remainder of his tour. His strikingly marked P-38 never revealed to his comrades that the cryptic MADU referred to his mother.

9. 432nd FS. P-38J-5 42-67290
Pilot: Hanson Crew Chief: A. J. Pinkerton
This was another 475th camouflaged P-38J. Records indicate that it was used during January and February 1944, and that it probably survived until all J-5 models were retired around mid-year.

10. 432nd FS. P-38L-5 44-25432
Pilot: Forster Crew Chief: J. R. Morrow/P.A. Harvey
This was the last regularly assigned P-38 used by Joe Forster before he went home on TDY late in March 1945. This P-38 seems to have remained relatively pristine until it was retired around the end of the war.

11. 433rd FS. P-38J-15 42-104305
Pilot: Lewis Crew Chief: G. Rath
This was one of the first natural metal P-38s to be accepted by the 433rd early in 1944. 433rd groundcrew veterans remember that the P-38 had an unusually long service life, lasting from the beginning of 1944 until it was sent to a service squadron in October of that year. Lewis remembers flying this plane most of the time that he was commander of the 433rd.

12. 433rd FS. P-38L-5 44-25930
Pilot: Purdy Crew Chief: Unknown
John Purdy was downed twice within a single month, once in the 433rd C.O.'s aircraft on December 11, 1944 and again in LIZZIE V on January 9, 1945. He apparently got the last of his air victories in this P-38 on December 17, 1945. LIZZIE was named after his wife, and many of the earlier versions were lost in action.

13. HQ, 475th/433rd FS. P-38J-15 42-104024
Pilot: MacDonald Crew Chief: Dietz
Colonel MacDonald's first natural metal PUTT-PUTT MARU was maintained in the 433rd Fighter Squadron from about April 1944 through at least August 1944. Some reports unofficially claim that this P-38 was wiped out in a bad landing on Biak, but Col MacDonald doesn't recall any such incident before the January 1945 landing accident.

14. 431st FS. P-38J-15 44-24155
Pilot: McGuire Crew Chief: F. Kish
Tom McGuire got his final PUDGY sometime in October or very early November 1944. The last days of 1944 were especially active for the hard-driving ace and he used up the airplane in less than two months. PUDGY V was awaiting transfer to a service squadron about the time that McGuire was killed.

15. HQ, 475th FG. P-38L-5 44-25643
Pilot: Loisel Crew Chief: Unknown
John Loisel doesn't remember this P-38 very well, suggesting that it was assigned to him but that he flew anything that was available. It is likely that this P-38 was maintained for him from about February 1945 until he took over the 475th after Colonel MacDonald went home in July 1945.

16. 432nd FS. P-38L-5 44-25600
Pilot: Summer Crew Chief: Unknown
This P-38 was probably used at the very end of the war, considering the 432nd emblem placed on the gun bay. By December 1945 another P-38L-5 (s/n 44-26310) was sporting the commander's number 140. Elliot Summer's scoreboard is maintained on the nose.

9

10

11

12

Several times the boat stopped and Hasty was struck each time. It occurred to him that they would dock at some native village where the Japanese would beat this helpless American pilot to show who was actually in power.

When the boat finally did reach Babo the first person to talk to Hasty when the blindfold was removed was a Japanese Naval officer who spoke fluent English and explained that he went to school in California. He told Hasty he would be safe in the hospital if American war planes respected the Red Cross. While the Japanese officer removed all of Hasty's insignia and other items on his person, the American asked about medical treatment for the burns on his legs that he suffered during the bailout.

The Japanese officer replied that medical supplies were short and that none could be spared. Hasty's right leg was especially badly burned and gave him a considerable amount of grief during the next few days of captivity.

That first night he spent tied up and miserable in a bare room. He heard the motors of a B-25 and a number of P-38s the next day. The officer jibed to him that his friends were trying to find him. Unknown to both of them, eight P-38s of Hasty's 433rd Squadron were covering the B-25 in search of

their missing friend. The Japanese may or may not have known that Lieutenant Paul Peters of the 433rd shot down one of two Oscars attacking the B-25. The Oscar crashed in flames about four miles from the crash site of Hasty's P-38.

Hasty's misery continue throughout the next days. His wounds were painful and his food consisted of about two bowls of rice and boiled water per day. On June 9 a raid by B-25s "plastered the hell out of that place", and Hasty began to think, ". . . school was out for me at the hands of my bomber friends. Bombs exploded close by and blast fragments tore out a corner of the hovel where he was tied.

Somehow he survived and the next day was taken at the first sign of light and thrown into the rear of a twin-engine bomber or transport. A cold sweat ran down his brow as he envisioned his chances if the Japanese plane ran into a P-38 patrol.

In the sequence of things Hasty survived the flight and landed at a base in Borneo. He stayed at an Australian POW camp where his allies probably saved his right leg with an application of pure Red Cross iodine. Within two days Hasty would be blindfolded, handcuffed and led away by a guard to the Philippines and Taiwan by transport plane before reaching

Another 433rd P-38 takes a banging during a training or operational accident at either Nadzab or Hollandia. (Krane files)

432nd pilot Dean Olson. (475th History)

Bob Crosswait with a Headquarters P-38 behind him. He got one victory over Hollandia for the 431st Squadron, but was killed in action at the end of June 1944.

Japan. He would be the first of the Satan's Angels to reach the enemy empire, but not under the circumstances he had anticipated.

Air Battle over Jefman

With the invasion of the Marianas in June the Japanese were stressed to the limit between the two prongs of the Central and Southwest Pacific. Attacks on Biak slackened when Japanese attention was diverted to the bombardment and storming of Saipan between June 13 and 15.

Fifth Air Force commanders were eager to crush Japanese air strength on Jefman, especially, as a prelude to taking strategic islands north of New Guinea. The first clear chance came on June 16, 1944.

The 475th could send up only thirty-eight P-38s early in the morning to land and regroup at Wakde. Unfortunately, even this relatively small number of P-38s was diminished when six of the Hades Squadron formation turned back sometime between the flight to Wakde and the mission itself. Colonel MacDonald was one of the six who SNAFUed his fighter was having engine trouble as he turned over leadership of the 431st to Major McGuire.

McGuire was happy enough to lead Hades Squadron off Wakde at nine thirty and proceed to the target, after rendezvousing with the B-25s over Roon Island. Even though the squadron had been reduced to nine P-38s by the time it was represented by some of the most pugnacious of Satan's Angels pilots. McGuire's wingman was a new pilot, Lieutenant Enrique Provencio and the second element of McGuire's Hades Red Flight was led by Bill Gronemeyer with George Veit as his wingman.

A makeshift second flight was created from the remains of Hades Blue and White with Captain Bill O'Brien as leader and Frank Monk flying number three position. Paul Morriss trailed along as fifth man in the flight.

Hades Squadron arrived with the bombers over the target at about 6,000 feet at forty-five minutes past noon. Within minutes a call came over the radio that enemy aircraft were taking off from an airstrip below. McGuire ordered external tanks dropped and led his fighters down to the attack.

The P-38s went down quickly from altitude to deck level with McGuire diving faster than anyone else. He made a pass at a Ki-51 Sonia and managed to get a glimpse of smoke pouring out of its engine before excess speed forced him to overshoot the target. Bill Gronemeyer was right behind the action and was able to shoot the Japanese plane down.

Several other Sonias were in the area, so McGuire picked one out and shot it into the water with a sixty degree deflection shot that ended up right behind the enemy plane. Lieutenant Provencio stayed on McGuire's wing during the action and saw not only the Sonia crash, but also some Oscars get off the ground to oppose the Americans.

One of the Oscars had latched onto a B-25 and McGuire led Provencio into a pass on the interceptor. McGuire actually stayed on the tail of his very maneuverable foe through several turns and finally exploded him on a reverse turn. Another Oscar was on the tail of several P-38s and McGuire got some good strikes on it before losing sight of the thing. Provencio was able to make a pass on this damaged Japanese fighter and fired a long burst that blew it up.

Captain O'Brien was leading Hades White Flight down in a great spiral and lost contact with McGuire's flight. He looked back to see that all four P-38s were still with him, then made a pass at a group of Oscars that appeared below. O'Brien fired at one of the enemy and was surprised to hear Frank Monk exclaim over the radio that the Oscar was going down with fire coming from the engine and cockpit; there wasn't much time for more than a brief shot at a passing target before O'Brien was forced to break away and look for another enemy plane.

One of the Oscars that O'Brien got a short burst at before it maneuvered out of his line of fire became an easy target for Lieutenant Horace "Bo" Reeves. The Oscar reversed its course at about three hundred feet off the water and turned sharply to the right. Reeves fired a 90 degree deflection shot for as long as he could hold the lead and noticed several small explosions around the wings and cockpit. The Oscar glided into the sea and bounced into the air for a few seconds before settling back into the water.

After he had witnessed O'Brien's Oscar crash, Monk made a diving turn to the left and saw two Oscars climbing up to attack Hades White Leader. Turning sharply into the path of the leading Oscar he fired a short burst before breaking away to avoid colliding with his target. Bo Reeves saw this Oscar fall in flames into the sea. Both Reeves and Monk saw

Morriss shoot up another Oscar which crashed in flames along the shore.

Howard Hedrick, who had been involved in Clover Squadron's combat, came along just in time to witness Morriss shoot down the Oscar, joining the Hades formation at the same time. Clover was being led on this mission by Major Meryl Smith of Headquarters. Between twenty to thirty Zeros, Oscars and one fixed-gear aircraft that was identified as a Val, but more likely was another Sonia, were sighted by the squadron and Smith led his flight down to attack and Oscar and the Sonia.

Smith quickly shot down the Sonia and his wingman, Lieutenant Bob Hadley, got the Oscar. Two Japanese fighters tried to get immediate revenge by jumping the American victors and managed to put several cannon shells and bullets into Hadley's P-38's wings and coolant radiator. Hadley responded quickly by getting a high angle deflection shot that destroyed one of the Zeros.

Possum Squadron had a brief but furious battle when it encountered twelve to fifteen Zeros, Oscars and Tojos over the battle area. Major Oliver McAfee was flying with the

Lent, McGuire, MacDonald and Loisel at Hollandia. They were the top scoring pilots of the 475th at the time with more than fifty kills between them by the end of June 1944. (Cooper)

MacDonald and his Headquarters staff. (Krane collection)

squadron and ran afoul of a Tojo that managed to damage his Lightning when he overshot the Japanese fighter. His wingman, Lieutenant Bert Roberts, was in position and shot down the Tojo before it could finish McAfee.

Roberts escorted his leader's damaged P-38 to cloud cover and rejoined the fight as part of another 433rd formation. C.J. Rieman and Joe Price accounted for two more Oscars on Possum Squadron's ledger, but Howard Stiles was last seen chasing an Oscar low-over the water of Jefman Harbor.

The antiaircraft fire coming from the ground and from warships in the harbor was moderately heavy, and some pilots believed that aerial bombs were used by the enemy during this mission. Beyond that, the interception was uncoordinated and rarely pressed home with any determination. The Satan's Angels score now stood at over 350 confirmed kills, with fewer than fifty pilots lost from all causes. Tom McGuire was now one of the highest scoring fighter aces of the Fifth Air Force with twenty confirmed kills.

Winding Down On Vogelkop

For the remainder of June 1944 there was little in the way of air action for V Fighter Command. The most obvious enemy presence was the barge traffic and occasional supply ship that ventured into range of Fifth Air Force bombers. Early in July Noemfoor would be invaded and the entire Geelvink Bay area would be closed to the Japanese. Once again, the method of island-hopping adopted by the planners of the Pacific strategy would strand thousands of first-line Japanese troops.

The Satan's Angels returned to Jefman on June 17 with another B-25 strike comprised of 38th and 345th Group bombers. The 38th Group sank or damaged a number of ships in the harbor while the 345th swept the nearly deserted airfields. Some 475th P-38s got into the action by strafing some barges and luggers near Cape Sorong. Japanese antiaircraft fire was so light and inaccurate that only one B-25 reported any significant damage.

Awards ceremony, probably at Hollandia. SSgt George Heap is standing almost at attention at the right corner of the formation. (Heap via Maxwell)

For the next few days the missions were extremely routine for the 475th. Aside from the A-20 and B-25 escorts or an occasional PBY (OA-10) cover mission, the P-38s ran into more danger just taking off or landing on their own bases.

Around June 26 the 475th began escorting a new attack plane on the A-20 missions. The A-26 Invader was meant to be a replacement for the aging A-20 Havoc and proved to be an impressive warplane. It was bigger than the Havoc and consequently carried a greater bombload over a greater distance. In addition, it was much faster than the Havoc, but none of the advantages impressed General Kenney who thought the A-20 was well-suited to his needs and looked with disfavor on the prospect of an involved conversion to the new attack type.

Lieutenant "Andy" Anderson went along on one of the early A-26 escorts and took positive notice of the new aircraft.

When Possum Squadron made rendezvous with the A-26s Anderson was surprised to find that he had to throttle forward to stay with the bombers. At low altitude with throttles pressed forward it seemed that the P-38 was hard pressed just to stay abreast of the A-26.

On the last day of June the 432nd took off late in the morning for its first divebombing mission. The target was personnel and supply dumps south of Kamiri drome. Only four bombs hit anywhere near the target and the mission was considered a failure. One civilian went along on the mission, but his bomb didn't do any damage to the Japanese, either. This was no ordinary civilian, but rather someone who would have a profound impact on the performance of the Satan's Angels as well as every other fighter unit in the South Pacific. He was Charles A. Lindbergh.

Chapter XI

THE LINDBERGH VISIT JULY-AUGUST 1944

Colonel MacDonald Meets Charles Lindbergh

On June 26, 1944, Colonel MacDonald had just finished a refreshing dip in the cold stream that flowed beside the 475th camp and was settling down to a checker game in his shack with newly-promoted Lt. Colonel Meryl Smith. It was mid-afternoon and the heat of the day wore off the effect of the cold bath rather quickly.

MacDonald was concentrating on the checker game, so the knock on his screen door was somewhat irritating. He growled for the interloper to enter and looked up to a rather tall and thin man with a balding forehead, who was in khaki uniform with no insignia. The man introduced himself, but Colonel MacDonald didn't quite catch the name when he and Smith stood up to shake hands with him.

The man took a seat while MacDonald and Smith resumed their checker game. He explained that General Donald "Fighter" Hutchinson the 3rd Air Task Force commander, had sent him over to discuss P-38 combat operations. Smith deferred to MacDonald and went on with his studied attention to the checker game. MacDonald took the stranger's question, but became somewhat peeved at losing four men off his side of the board because of his divided efforts.

Subsequent questions by this civilian were unusually intelligent and very technical. MacDonald took a harder look at what he thought was just another stateside expert who would gather some vital information before he simply disappeared. He asked if the man were a pilot, to which question the answer was yes. After a bit of startling recognition dawned the colonel asked if the man could be the Charles Lindbergh. He answered that that was indeed his name.

The conversation drew more attention from both MacDonald and Smith and they abandoned their checker game. the aviation bull session went on into the early evening hours and attracted other curious pilots including Tom McGuire. Probably never before had such a distinguished company gathered to discuss the merits of a single fighter aircraft type in this case the P-38.

It was finally decided to do the only thing that would show Lindbergh what these P-38 aces were talking about; namely, take him up for an operational flight. McGuire agreed to take him on his own wing as protection for Lindbergh. So enthusiastic was McGuire about flying the mission that he offered bits of useful information like taking along extra chocolate bars to endure the 600 mile trip to the target area.

Later, when Lindbergh had left in his jeep to return to General Hutchinson's quarters, MacDonald and Smith realized that they were going to take responsibility for letting a civilian in his forties go along on a combat mission. The two fighter veterans knew that only very young pilots with a good deal of training stood a reasonable chance in an air battle.

Meryl Smith, Charles Lindbergh and Charles MacDonald at Hollandia before the move to Biak. (Gregg)

The two men jumped in their own jeep and ran over the bumpy ground to Hutchinson's quarters. A visit to the General was overdue for both officers and he was glad to spend the evening talking with both of them and Lindbergh about Lindbergh's mission to learn more about the successes of the P-38 twin-engine fighter. Lindbergh was in the SWPA as a representative of United Aircraft Corporation, and had already begun making valuable suggestions regarding the F4U Corsair which was being manufactured by his company.

Smith suggested when the conversation was breaking up that it would be better if Lindbergh stayed in the 475th camp if he were going on the morning mission. Everyone agreed and Lindbergh began a two month stay with the Satan's Angels, his longest with any air unit in the Pacific.

The Lone Eagle in Combat

It was raining at 5:30 in the morning when Lindbergh got up and started to prepare for the mission. He had checked out in

the P-38 a short time before. Mel Allan was on hand with a group of excited groundcrewmen when Lindbergh appeared on the flightline with Colonel MacDonald and Smith and Major McGuire. The four pilots were due to take off at about 10:30 that morning and everybody wanted to be sure that things were perfect, so a general fuss was made over Lindbergh's aircraft and equipment.

One of the things that Sergeant Allan, among the other men who were in the crews, advised Lindbergh was to replace the CO_2 bottle that he had removed from his seat pack rubber life raft. It seems that Lindbergh hated to take along any added weight on a flight and thought that he could easily inflate the raft with his own wind. The NCOs convinced him that he could very well be injured or otherwise weakened in a crashlanding at sea and the quick-inflating feature of the bottle could save his life.

Lindbergh grinned his famous smile at the obvious concern for his safety. He quickly acknowledged the wisdom of the veteran groundcrews and a CO_2 bottle was promptly attached to the collapsed rubber dinghy.

When the flight did take off over the "green-carpeted hills, plains of kunai grass and an occasional native village . . .", as Colonel MacDonald described the beginning of the mission, Lindbergh slipped easily in on the wing of Major McGuire. MacDonald noticed that Lindbergh flew a perfect formation on McGuire's wing. It was even more impressive that he flew so well considering the fact that he had checked out in the P-38 only a few days before with Colonel Bob Morrissey's 49th Fighter Group and had very few flights in the type with the 49th and 8th Fighter Groups, and had flown up to Hollandia in Dick Bong's old P-38.

MacDonald and Lindbergh by the second P-38 PUTT PUTT MARU. It was claimed that this P-38 was written off in a bad landing on Biak sometime in July or August 1944 and MacDonald got his first P-38L, serial 44-24843. (Cooper)

C-47s arriving at Hollandia to transport the 475th air echelon to Biak. The move lasted from July 10 to July 14, 1944. (Hanks)

433rd P-38s on Boroka Strip, Biak. Note that the individual aircraft number has moved from the nose to engine cowling, July-August 1944. (Anderson)

McGuire, Lindbergh, Smith and MacDonald on Biak.

The flight sighted some Japanese ships around Sorong and had to dodge some of the inevitable flak. Later, along the rugged northern coast of the Vogelkop, there were discovered some barges and luggers camouflaged with leaves and branches. Lindbergh took his place in the strafing circle and fired three bursts; the first two missed, but the third sent tracers into the target which bounced and danced with the impact.

Another target for Lindbergh farther along the coast took some hits above the waterline and began to flame just as the P-38 was passing overhead. The lugger's fuel tanks exploded an instant after Lindbergh's fighter cleared the ships mast by about thirty feet. After a few more passes only one of Lindbergh's .50 caliber guns would fire a few erratic bursts, so he was through with his first day of combat.

Lindbergh Learns His Lessons

For the next few days Lindbergh was in the air for every combat mission he could possibly fly. On July 1 he and Colonel MacDonald led a 433rd flight on a sweep of the Geelvink Bay coastline to Nabire and the Japanese airstrips between there and Sagan to the northwest. Lindbergh was especially eager to encounter Japanese fighters to fully understand the implications of combat with the lighter and more nimble enemy aircraft compared to the more powerful but heavier twin-engine P-38.

Possum Special Three was Lindbergh's designation on the flight and he cheerfully led his wingman low over the treetops and into the valley of a river near McCluer Gulf. There were aircraft visible on the fields at Sagan and Lindbergh went down for a strafing pass on Otawiri strip. He blasted one hut, but held the sights on his target so long that he nearly flew into the ground and cleared the hut by no more than fifteen feet.

Japanese light flak was in abundance, although very few hits were scored on the P-38s as they passed over the field. The four pilots noticed that the parked aircraft on the side of the field were wooden dummies, designed to entice marauding enemy planes into their fire.

After circling the Geelvink coastal area, at one point close to the Babo area where Bill Hasty had been held until a few weeks before, the Possum flight headed home. Lindbergh had acquired just a bit more data from his combat experience.

The next day Noemfoor Island was invaded by American troops and Lindbergh again flew with the 433rd as Possum White One. It was hoped that the Japanese would challenge the landings in the air, but the Possum Squadron patrol was uneventful and Possum White landed early because its relief showed up ahead of schedule.

On July 3 Lindbergh was Possum White One on his first escort mission. Once again the B-24s were scheduled to hit Jefman and it was hoped that Japanese interceptors would appear. As it happened, there was no aerial opposition and the mission turned into a routine flight.

The twelve P-38s of Possum Squadron tried to make the most of a dull trip by strafing barges that seemed to grow out of every bay and inlet along the Vogelkop Peninsula. After a time, one of the P-38s reported its fuel was low, so Colonel MacDonald ordered it back to base. A few minutes later another called out the same report and headed back. During one of the runs on a target Lindbergh noticed that his own wingman was flying in circles overhead and found out by radio that the man was also low on fuel.

Finding out that the other P-38 had "About 175 gallons", Lindbergh told him to reduce rpms on his engine to 1,600, put his mixture control in auto-lean, and to open throttle enough to stay in loose formation. The two planes set course for Owi Island, and when they landed the other pilot was amazed to find that he had seventy gallons of fuel left. Lindbergh astounded him even more when it was discovered that his own tanks had 260 gallons.

When the word got around that Lindbergh could save that much fuel on long missions he became more than just a visiting celebrity. At first there was skepticism that his range-extending techniques would simply foul the sparkplugs and damage engine manifolds. In the course of events it was discovered that no significant harm was done to the engines, but many crewchiefs cursed the endless task of changing plugs that attended the Lindbergh process.

It was not a new idea that the Lone Eagle was expounding. Most veteran pilots knew that reducing rpm and manifold pressure gave them considerably longer time in the air. The question was applying the procedure to a powerful engine like

#184 was the usual P-38 flown by Lindbergh with the 431st Squadron.

the Allison V-1710 that powered the P-38. Lindbergh was a careful engineer who studied and applied the proper numbers for maximum engine efficiency.

Someone in a high place to the south in Australia learned of Lindbergh's participation in combat and sent word to Colonel Morrissey to end it. Morrissey informed Lindbergh of the decision on July 5 and the next day Lindbergh was off to Fifth Air Force Headquarters. He spent several days with personage like General Kenney and General MacArthur, convincing them that his time with the 475th would eventually result in 700 mile ranges for the P-38 as well as substantial increases for other fighter types. He would return in success to the 475th within a few weeks.

Beautiful Biak

Within a day or two of Lindbergh's departure south the 475th began tearing down its campsite at Hollandia for the movement to Biak, just north of Geelvink Bay. On July 10 the first

Left: Lindbergh in the cockpit ready for flight. Note the tail of Pudgy III in the background.

Above and opposite: Lindbergh and McGuire briefing before or after a mission.

echelon of the group flew to the island. The 432nd Squadron didn't get off on the first air move because of bad weather that crept in, but flew one of its last missions from Hollandia on the eleventh when twelve of its P-38s escorted A-20s to Kokas.

On July 12 twenty transports loaded with men and equipment set sail for Biak. Mel Allan was on one of the transports and remembered the hold of the ship as a "dark, stinking mess." The ventilating fans did not adequately cool the interior of the ship and the men sat miserably in their own sweat within the confines of their crowded quarters.

Allan found what seemed to be a laundry room with large tubs and spigots that tapped into fresh water. When everyone else was at chow, he slipped into the laundry room with soap and a towel and had a refreshing cool bath that removed all the accumulated sweat and salt. He was not quick enough getting out, however, for some of the men returning from supper saw him and decided to do the same thing. Before long the fresh water reservoir for that part of the ship was probably being depleted because the precious liquid was shut off on that part of the ship.

The ships unloaded on a Biak beach on July 14. Most of the heavier pieces of equipment like bomb trailers and fuel trucks had been turned over to service squadrons to facilitate the move, and most of the heavy tools were aboard ships that would have to wait to be unloaded.

Biak itself, at least on the point where the 475th personnel were unloaded, was hard coral with underbrush about ten to twenty feet high. The reflection of the coral was dazzling in the tropical sun and was nearly unbearable in the noonday heat, but the surface was very hard and P-38s were quickly brought in on the ready-made hardstands.

Stubborn Japanese resistance was being experienced nearby and G.I. foxholes were almost within the 475th peri-

meter of the 475th camp. Intermittent firing could be heard coming from nearby and desperate Japanese air raids were a common nighttime danger. 475th personnel were warned not to venture out at night, but one man had an overwhelming urge on one occasion and ventured out. He had the bad luck to trip one of the American tripwire grenades and was lucky enough to dive out of the way before alerted guards sprayed the area with gunfire. The only sound when the shooting stopped was the unlucky man cursing his wet pants.

Other dangers on Biak came from unexpected sources. By the end of July there were a number of B-24 and B-25 units on the island's three airfields as well as an Australian P-40 unit and some F-5 photo reconnaissance Lightnings. One of the F-5s, in fact, was almost the cause of a number of fatalities among the ranks of the 431st Squadron groundcrew.

Sergeant J.D. Barrow and his crew were working on his P-38 during the afternoon of July 29 with about a half-dozen other 431st enlisted men, including Mel Allan, standing nearby the four P-38s on that part of the revetment. Unknown to them at the time, an F-5 with Colonel Lehman of the 7th Service Group was just taking off from the Mokmer Strip and heading in their direction.

Barrow and his men suddenly began scurrying away as if a Japanese air raid had appeared overhead without warning. Allan watched in dumb detachment as the F-5, ". . . pulled up rather steeply, banked in a steep climb, curled over and headed right down towards us in a full speed dive."

The F-5 crashed fifty feet away from Allan and smashed into the tail of Barrow's aircraft. There was no explosion which would have killed everybody in the vicinity but the F-5 bounded over the P-38 and dragged it into the second aircraft on the line.

Operations shack data gathering.

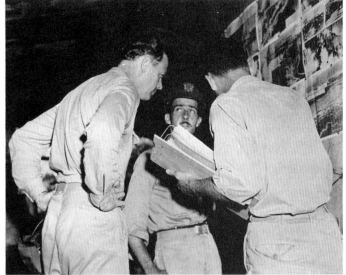

All three planes ended up in a heap that eventually started to burn before the horrified eyes of everyone there. It was later conjectured that the pilot had suffered a heart attack just as the F-5 became airborne and may have been dead before the crash.

Lindbergh Teaches His Lessons

MacArthur and Kenney had been persuaded to allow Lindbergh back on combat missions. He had convinced them that the more he learned of actual operational problems, the more he could pass on in terms of his expertise. On July 16 he landed in a P-47 that he agreed to ferry to Owi Island.

Four days later Lindbergh and Colonel Morrissey were waiting in the morning at the end of the strip in their P-38s for Colonels MacDonald and Smith. At precisely 10:00 the pair of them buzzed their waiting comrades on the strip below, and Lindbergh and Morrissey were off to join them.

The sweep covered the southern part of the bird's beak on the southwestern part of the Vogelkop. Several barges and buildings were strafed by MacDonald and Smith, but no other significant action was encountered from the arc of the flight that covered the distance from the Pisang Islands to Fakfak, and as far south as Ceram.

By the latter part of July the value of Lindbergh's lessons on fuel economy was beginning to take effect in the 475th. Weather was always an unknown factor in New Guinea air navigation and some operational losses were attributed to this cause. For example, on July 28 Lieutenant Bill Elliot was returning from Noemfoor to Biak when he radioed that his P-38 had only five minutes of fuel left and that he was going to ditch into the sea. The weather had been troublesome all along the route and may have been the cause of Elliot's forced landing and loss.

Other pilots found that Lindbergh's technique gave them much more latitude in skirting or enduring adverse weather. More pilots began to realize that the P-38 could overcome virtually all obstacles that weather could present, the twin-engine fighter seemingly possessing an unlimited flight endurance. Undoubtedly, fewer of the group's P-38s would have been lost on Black Sunday if the fuel-saving procedures had been implemented earlier.

From all accounts, it is clear that the pilots of the 475th took Lindbergh to their hearts and considered him one of their

An informal Andy Anderson. He always valued the fact that he was on the July 28, 1944 mission. (Anderson)

own. In return he considered them rough hewn, for the most part, and criticized them for their shock-troop appearance, bawdy decorations on their P-38s and especially for their brutal attitude toward the enemy.

One of the pilots that Lindbergh spent a great deal of time with was Major Tom McGuire. Much has been made in written accounts of the fact that McGuire persistently taunted Lindbergh and made him do small favors, knowing full well that Lindbergh could not effectively retaliate because of his tenuous guest status. Other 475th people claim that the bantering between the two men was mutual and generally good natured.

Lindbergh was an incorrigible practical joker who enjoyed putting a mild stunt to the unsuspecting. One rumored turnabout that he is supposed to have dealt to McGuire is said to have happened when McGuire casually requested him to retrieve the ace's famous battered service hat. Lindbergh apparently hastened off and later returned with a second lieutenant's flight cap about three sizes too big.

Lindbergh's Aerial Victory

The first strike to the Halmaheras was scheduled for July 25. Lindbergh was especially eager to fly the mission, since it would be his first really long range attempt. However, Colonel MacDonald came back from Owi Island with instructions from Colonel Merian Cooper that Lindbergh would not be allowed to go on the mission.

Lindbergh was puzzled and a little incensed by the apparent waffling on the part of the higher authorities. Both MacArthur's Headquarters and Kenney's Far East Air Force had agreed that he could provide valuable service from his combat experiences. Why was there apparently change of policy from moment to moment?

It mattered little what the powers that be dictated on July 25 because the mission was scrubbed in spite of a call from Colonel Morrissey authorizing Lindbergh's participation. Two days later the big show to the Halmaheras did get off with fifty-two B-24s covered by about fifty P-38s from Satan's Angels as well as two squadrons of P-38s from the 8th Fighter Group. The results were satisfactory with fifteen Japanese aircraft shot down and a similar number destroyed on the ground. Lieutenant Leo Blakely, who had only recently joined the 432nd Squadron on July 11, was the only casualty when he crashlanded near Cape Waios and was safely picked up by a PBY.

Lindbergh was pleased with his part in the mission and eagerly awaited the next day's return to the same area. He had flown on July 27 with the 433rd Squadron which had covered the B-25 strafers and, thus, had not engaged any Japanese aircraft. The next mission would be different.

The day was cold and dark at 5:30 on the morning of July 28, 1944 when the pilots of the 433rd Fighter Squadron rolled slowly out of bed. After breakfast the dawn was coming in gray and cold while the P-38s blurped and crackled into life. The B-25s had already taken off for the south of Ceram, the 433rd's target would be the island airstrips of Haroekoe, Liang and Ambon, all just south of Ceram.

Seventeen Possum Squadron P-38s would take off at 7:40 with Warren Lewis leading as Possum Red One. Lieutenant "Andy" Anderson was his wingman in the hope that Andy would finally get to fire his guns at a Japanese aircraft. Herb Cochran flew element lead and Bert Roberts was Possum Red Four to bring up the rear of the flight.

Colonel MacDonald led Possum Blue Flight with Captain

Ed "Fishkiller" Miller by Chase Brenizer's #182 on Biak. Miller got his name when one of his bombs fell short on a mission to Nabire, and hordes of fish rose to the surface of the explosion vicinity. (Anderson)

Herb Cochran took a dunking on the day that Lindbergh got his aerial victory, and had to paddle his way back to camp.

McGuire and Lindbergh again. The two apparently got along rather well, since McGuire's tendency to provoke anger was matched by Lindbergh's love of practical jokes.

Danforth Miller as his wingman. Lindbergh was Possum Blue Three with Lieutenant Edward "Fishkiller" Miller as his wingman. Possum Squadron kept climbing and Lindbergh put on his oxygen mask at 15,000 feet. The clouds gradually began to close in while the P-38s rendezvoused with the 345th Bomb Group B-25s. At last, the lower flying B-25s radio that they can't get through the rain and clouds and are turning back.

Nobody had noticed that Possum Red Three had disappeared from the flight. Herb Cochran's P-38 had developed engine trouble while he was trying to form up with the lead element and he fell behind Red Flight. His wingman saw him swing away and merely thought that he was aborting the mission.

Both of Cochran's engines had quit and he was forced to crashland into the sea just off Sorido village. Quickly getting into his life raft, Cochran waited for the PBY to rescue him.

What came first was the returning force of B-25s which happened to jettison its bombload right over his head!

Frantically, Cochran waved his oar and then tried to paddle out of the way. Fortunately, the bombs missed him, but he rowed his little boat all the way across Sorido Lagoon and walked back to the 475th camp, shaken but unhurt.

Meanwhile, MacDonald had used his prerogative as group commander to initiate an impromptu fighter sweep. He called Possum Yellow One, veteran Captain Jack Fisk, to join him and the two flights ranged at about 13,000 feet over the south of Ceram. The interior of Ceram is a solid carpet of clouds, but the southern coast is clearing into good weather with just a few cumulus at about 4,000 feet.

Leading a formation of eight P-38s, MacDonald was in his element; the flight was small enough to maneuver freely over the most promising territory and yet large enough to take on any enemy air force it was likely to meet. The two Possum flights flew up and down the south coast of Ceram on the alert for Japanese aircraft.

Over Amahai, on the south central coast of Ceram in the Elpapoetih Bay, some black puffs of flak came up to the P-38 level and MacDonald split the two flights, turning out toward Amboina with his own Blue Flight. All the American pilots are becoming edgy with the possibility of running into the enemy or worse, not running into him when radio calls are heard from Captive Flights (9th Fighter Squadron call sign) that they are engaging Japanese aircraft over the bay.

Unknown to the Americans at the time, two Ki-51 Sonia reconnaissance aircraft were on a search and rescue mission for two of their own downed planes. Captain Saburo Shimada, commanding officer of the 73rd Independent Chutai, and Captain Fukumichi Oda of the same unit had run afoul of the 9th Fighter Squadron. The veteran P-38 pilots had claimed one aircraft shot down, but were having the devil of a time with another that outflew them at every turn.

The pilots of the 9th's parent 49th Fighter Group and those of the Satan's Angels felt a sometimes heated rivalry for the honor of being called the top fighter group in the Fifth Air Force. Many of the 475th's great leaders came from the 49th and the younger unit had claimed many incoming pilots and P-38s; the 9th Squadron had only recently re-converted back to P-38s after using P-47s for a time in deference to the needs of the 475th.

Thus, it was not with a great deal of elation that they watched the P-38s of Possum Blue Flight appear overhead. But the Sonia that has eluded eight P-38s of Captive Squadron is now fair game for the new flight approaching partly because many of the 9th Squadron P-38s are running out of ammunition.

MacDonald makes the first run on the enemy plane after he orders tanks dropped and the four gun switches of possum Blue Flight are turned on. A full deflection shot from MacDonald's P-38 draws smoke from the Sonia, forcing it into a sharp opposite bank. Danforth Miller get a brief shot, but the Sonia continues to turn in the direction of Lindbergh's P-38.

Lindbergh fires his own guns and sees strikes on the fuselage of the Sonia. Amazingly, the Japanese plane seems to shed off all damage and continues to turn into Lindbergh's P-38. Lindbergh sights on the engine of his target and holds the trigger down. When the two aircraft are dangerously close, Lindbergh pulls back on the wheel and can see the cylinders of the Sonia's engine before it passes under. An uncomfortable shudder engulfs the P-38 from the Sonia which misses by no more than ten feet.

Only P-38s now are seen in the sky. Apparently, the Sonia pilot was wounded or the Japanese plane's controls were shot away because it fell into a spiral that tightened into a wild spin that took it down into the bay. Only a frothy white circle remains on the sea below.

Lindbergh's wingman stayed with him and witnessed the Sonia blazing away at Lindbergh before it fell out of control and crashed. The rest of Possum Blue has mysteriously vanished, possibly above a thin cloud layer above. Radio reception is poor and Miller is also lost while they climb through the cloud layer.

Lindbergh set course for home and finally makes contact with all three members of Possum Blue Flight. They join up and finally land at Mokmer about five minutes before four in the afternoon. The mission has taken nearly eight hours.

In all probability, the Sonia that Lindbergh shot down was piloted by Chutai commander Shimada. The 433rd Squadron final mission report for number 3-407 makes the remark that the "Enemy pilot seemed experienced and skilled, and used plane's maneuverability to advantage." It was a skilled and cool pilot that Lindbergh faced, who managed to cope with at least a dozen P-38s before he succumbed to one of the great American pilots of all time.

Night Bombing

Throughout the end of July and into August the Japanese continued to send small bombing raids to harass and perhaps damage American forces on Biak. The threat of these raids subsided after a number of the bomber bases were attacked and P-61 Black Widow nightfighters arrived on Biak.

One of the raids could have had more than a just a bit of harassment effect when it came at an inconvenient time for the 433rd Squadron's comfort. Mostly, the raids came late at night after everybody had a chance to use the toilet facilities, but one night the bombers came early.

Clarence Rieman had just filed out of the screened in five holer that overlooked the beautiful sea toward Owi Island when the bombers came over. He noticed that everybody seemed to have made it out and taken cover before the bombs began to fall. When the string of bright flashes ended and the bombing seemed to have stopped, 433rd Commander Warren Lewis emerged from the facility and Rieman heard him remark casually, "Well, my problem was cured in a hurry."

Lindbergh's Last Combats

The last few nights of July were perhaps the most companionable that Lindbergh spent with the pilots of the 475th. On the evening of the day that he got his aerial victory Lindbergh

Bill O'Brien earlier in the war when he had only one victory. He was killed in a head on collision with a Tony Fighter near Ceram on August 4, 1944 in this P-38. (Krane collection)

Left boom view of O'Brien's P-38. (Krane collection)

gathered with MacDonald, Smith, "Pappy" Cline, John Loisel and a ukulele chorus for a general bull session.

After the session had broken up MacDonald mentioned to Lindbergh that the full moon was perfect for a bombing attack. Lindbergh was concerned about the box of TNT that he and

Edgar Childs, crewchief of CILLE. (Krane collection)

Colonel MacDonald kept as a roommate. The explosives were used to gather in fresh fish when the high-concussion packets were thrown from a liferaft like depth charges.

Early in the morning of August 1 MacDonald and Lindbergh went out to the revetment area where the turbos on the P-38s were still glowing red in the predawn darkness. Other 433rd pilots were lugging their parachutes behind them trying to read the numbers of the Lightnings to find each assigned aircraft.

The main mission was eventually scrubbed and Lindbergh persuaded MacDonald to fly a four-plane fighter sweep to the distant Palaus where Japanese fighter strength is reported to be 150 planes. By 9:30 the usual flight of MacDonald, Danforth Miller, Meryl Smith and Lindbergh was off for Babelthuap in the Palaus.

Peleliu one of the Palau islands that would be invaded by the Marines in September came into sight after MacDonald's dead reckoning and the four P-38s began their sweep around noon. within half an hour the crew of a sailing yacht are startled to see four American fighters skim over their sails and a 400-ton lugger is strafed and set afire.

When MacDonald was just preparing to turn onto another Japanese ship for a strafing run he noticed he noticed two enemy aircraft patrolling at one end of the convoy. At the same moment Smith called out, "Bandits two o'clock high!" Two Nakajima-built floatplane Zeros known by the codename Rufe were the newly identified targets and MacDonald urged caution over the radio even as he went into a fast, climbing turn to get on the tail of the nearest plane.

The Japanese under attack undoubtedly had seen his attackers and made for the clouds above. The other Rufe dived away to the south and was chased down to the water by Smith and Lindbergh.

MacDonald fired a three-second burst when he got into range of his Rufe. Both he and Miller saw the plane burst into flame at the wingroot and fall off into a spin toward the sea. After the Rufe hit the water, MacDonald turned around and looked for Smith and Lindbergh. He found them just in time to see Smith's kill skip off the water like a flat stone and slip again beneath the waves.

However, there was another enemy plane closing in on Smith's tail and MacDonald shouted a warning even as he dived to reach the Hamp first. Smith had rushed in ahead of Lindbergh to get the Rufe and consequently put himself in range of the new threat that MacDonald is able to discourage with the approach of his P-38.

MacDonald was only momentarily cheated of a victory when the Hamp pulled up into the clouds because, while he was reassembling the flight, a Japanese divebomber (Val, Judy or perhaps Sonia) put in an appearance, sauntering along over the ships below. Ordering the others to get above the clouds and watch for Japanese fighters, he got on the tail of the enemy and shot it down in flames with a single burst.

Now realizing that enemy fighters would be scrambling to check on the reported P-38 force, MacDonald looked at his own fuel supply and reassembled the flight once again for an indirect course home through the expedient cloud cover. Within a short time the worry about Japanese interception proved valid when another Zero came down through the clouds onto Smith's tail.

"Zero at six o'clock! Diving on us", cried Lindbergh over the radio. He tries to turn into the Zero to save Smith, but the turn is too soon and the Japanese has a better target in the P-38 that is now much closer. MacDonald shouts for Lindbergh to break right and the P-38 under attack obeys immediately.

There is nothing else for Lindbergh to do but obey and crouch behind his armor plate. He is too low to dive and the Zero is coming in much too fast for the P-38 to outrun or outclimb it.

The distance between MacDonald and Lindbergh closes rapidly enough for MacDonald to get off a good burst into the immediate flight path of the Zero. To get even a fast burst off MacDonald had to center his sights on Lindbergh's P-38, but Lindbergh is relieved to look back and see the Zero climbing away toward the clouds. Both Smith and Miller also get in bursts that draw smoke from the Zero before it disappears.

That evening MacDonald got a phone call from Colonel Morrissey suggesting that he was in trouble for the unauthorized long range mission. Lindbergh believed that he was to blame for the conception of the mission and that MacDonald was unfairly criticized. It seems that Fighter Command had been claiming their fighters could not cover the bombers as far

Ted Hanks taking advantage of a lull in activity on Biak. (Hanks)

as the Palaus and the four P-38 sweep had proved them wrong!

Whatever the political hot potato of Lindbergh's visit had finally brought on the 475th, MacDonald was finally vindicated on the grounds that he went by the letter of the law; every mission that Lindbergh flew was authorized. Nevertheless, MacDonald was grounded for sixty days and ordered back to the United States. This break in his combat tour did nothing to slow his magnificent record as commander of the 475th. Meryl Smith became interim commander of the group and MacDonald got to go home and see his baby son for the first time.

Lindbergh flew a strike mission to Liang strip on August 4, the same day that MacDonald left for home on leave. This time the flight was made up of eight P-38s from the 431st Squadron. Captain Bill O'Brien would be leading Hades Red Flight while Major Smith led Blue Flight with Lieutenant James Barnes as his wingman. Lindbergh would be Hades Blue Three with Lieutenant Bo Reeves as his wingman.

Somewhere on the way to the rendezvous there was a SNAFU and for a time Lindbergh was confused about the

Angus McMurchie and George Rath, the crew of Major Lewis's #170 until he went home in August 1944. (Hanks)

status of the mission. By the time he decided that the mission was still on, the 431st was out of sight and he joined up with the P-38s of the 9th Fighter Squadron.

Before the P-38s of the 9th Squadron with Lindbergh tagging along reached them, Hades Squadron made contact with two Japanese fighters. Hades was north of Liang and over Ceram in the vicinity of the Nala River at 21,000 feet when O'Brien initiated an attack on a Tony and another radial engine Japanese fighter about three thousand feet below. Owen Giertsen, recently returned to the 475th, was behind O'Brien and last saw him getting on the tail of the Tony.

By this time Lindbergh had arrived with the 9th Squadron and several Japanese aircraft were observed, prompting the jettisoning of external tanks. The sky is almost completely overcast, obscuring visual sightings. Lindbergh witnesses several Japanese aircraft heading for the clouds above with many P-38s on their tails. Out of the corner of his eye he also sees a bright flash; two planes have collided head on.

When he lands back at Mokmer strip Lindbergh learns some of the sketchy details of what he fears is the loss of a

friend. Captain Bill O'Brien was closing on the tail of the Tony when the Japanese suddenly pulled up in a tight loop. O'Brien's P-38 with squadron number 132 and the personal name "Cille" on the nose rammed the inverted Tony head on.

Incredibly, a parachute opened a few seconds later almost directly below the collision site. For several days the hope of finding O'Brien alive encouraged the 431st pilots and Lindbergh to fly search missions, but he was permanently listed as missing in action and joined the other Satan's Angels who were lost over the vast Pacific. The Tony was credited to him as his fourth air victory.

Lindbergh Leaves the 475th

After MacDonald left on the fourth of August (he went with Major Loisel, who was also going home for a leave and tour in the General Staff School) Lindbergh moved in with Tom McGuire. For the few remaining days of his visit it would be a better location for getting to the squadron ready tents.

On August 13 General Kenney sends down word that Lindbergh is not to do any more combat flying. The Lone Eagle has already started packing, so there is no disappointment. He flies over to Owi Island as a passenger in an L-5 and is greeted by Marine aces Joe Foss and Marion Carl. His next work assignment is to improve the efficiency of F4U Corsair divebombing missions, but his value to the fighters of the Southwest Pacific especially the P-38 is just beginning.

August Winddown

By August 15 an airstrip was operational on Middleburg Island off the coast of Cape Sansapor. With impending American operations in the Central Pacific the Japanese were being pressed out of the area east of the Philippines. The Vogelkop Peninsula was virtually in Allied control and fewer contacts with the Japanese were made.

The weather was a major factor in limiting operations. Between August 15 and the end of the month weather canceled six missions for the 432nd while five others either were completed to the primary target or diverted to secondaries because of the weather.

If the days were dull because of limited aerial activity the maintenance schedules were catching up with the demand and more P-38s were available on the Satan's Angels operational list. It was also a good time for training and new pilots such as Lieutenants Bob Shuh and Laurence Dowler got their orientation flights in the latter part of August.

The 432nd flew its last significant mission of the month on the 28th when sixteen Clover P-38s took off early in the morning, each plane carrying a single 1,000 pound bomb, to divebomb and strafe the Halong seaplane base. The fifteen P-38s that made it to the target area put all but one bomb in the hangar or barracks area. The other bomb landed in the water near what was believed to be a dummy float plane.

Belly tanks were then dropped and the strafing begun. Two barges were destroyed and another ship damaged. All P-38s returned to base and the Halong base on Ceram was left with numerous small fires on the runways and on ships in the harbor.

August 28 was also the day when the late June rotation list was supposed to get off. Since the beginning of the year morale had been high for the sunbaked enlisted men who had diligently done their duty for two years or more in New Guinea. As it happened, the transport planes didn't make it and the men suffered one more disappointment in their tropical service. Fortunately, it was the final letdown when the planes arrived the next day to take them home.

The last day of the month gave two 432nd officers a unique experience. Lieutenant Owen Griffith, the squadron Intelligence Officer, and Lieutenant Walter Weisfus were out exploring in a jeep when they stopped to examine a dead bird. They were about two miles beyond the 163rd Infantry Headquarters and should have been especially cautious.

This time the lesson was probably struck home with some impact. Griffith was startled to see a Japanese soldier hiding behind a tree and, because they were both unarmed, the Americans began some excited conversation. Fortunately, the Japanese decided to come out with his hands high, giving the 432nd its first captured enemy soldier.

Lt. Alan Sidnam's #172 on Mokmer Strip, Biak, August 1944. (Krane coll.)

Chapter XII

LONG-RANGE LIGHTNINGS SEPTEMBER-OCTOBER 1944

Homeward Bound

MSgt Ted Hanks was one of the enlisted men who was aboard the transport on August 29. He had been in New Guinea since 1942 and hadn't been so excited about a trip since he enlisted in October of 1940. He had never been happier in the service as he was servicing P-38s of the 433rd Fighter Squadron. By the time he left the squadron he was a flight chief over ten of the Lightnings.

Hanks joined other 475th enlisted men such as TSgt Wilson Howes and SSgt Carl Plecker of the 432nd at Finschhafen. There they boarded a Navy ship and headed for San Francisco to be once again on American soil within a few weeks.

During September a number of veteran Satan's Angels pilots were also rotated home. Clarence Rieman was one of those returning pilots, and happened to have the most accrued combat missions at 162. Squadronmate William Jeakle followed closely with 160 and Stanley Northrup was leaving with 139 missions to his credit.

Vincent Elliott of the 431st was leaving for home with 117 missions and seven Japanese aircraft to his credit. Charles Samms was returning with 96 missions. Cliff Mann of the 432nd Squadron had 121 missions; James Farris, 119; John Michener, 117 and Edward Dickey, 106 when they boarded the transport for home in September.

The rotation program would finally catch up to the squadrons in the Philippines. Starting from the end of 1944 and beginning of 1945 most of the old veterans of the 475th Group would return home. John Loisel would be a prominent exception to the rule when he stayed on throughout the war and took over the Satan's Angels in July 1945. It has been estimated that he served a longer tour in the Pacific than any other Far East Air Forces fighter pilot.

September Missions

A preview of sorts was granted to the 432nd Squadron when it participated in the first escorted bombing mission to the Philippines. On September 2, a force of B-24s, escorted by P-38s of all three V Fighter Command groups, struck at Davao, Mindanao. Five Clover Squadron planes took off from Middleburg Island and rendezvoused with the bombers over the target. The bombing was regarded to be excellent, but the Clover pilots were disappointed when they didn't see a single enemy fighter. Only some inaccurate flak came up to greet them and a pair of stray Japanese aircraft fell to other squadrons.

For the most part September was a month of scattered ground attack missions, and little air opposition was experienced by any Fifth Air Force crews. It proved to be one of the least active months in the history of the 475th Group.

433rd pilot Lt. Clarence B. Simons of Minneapolis, Minnesota on Biak. (Anderson)

That is not to say that the month of September was without its thrills. Lieutenant Bob Schuh had been with the 432nd for less than a month when he participated in the September 3 long-range escort to the Lembah Straits in the Celebes.

Schuh was hit by antiaircraft fire in the Lolobata-Miti Island part of the flight in to the target. His P-38J-5 was hit in the right boom and vertical stabilizer, the control cables and pulleys were damaged and the IFF box self-destructed. How-ever, though he was somewhat shaken, Schuh was uninjured and was escorted safely back to Biak.

On another long range mission to the west coast of Halmahera on September 10 Lieutenant Wesley Hulett of the 433rd Squadron had a similar experience when his P-38 was struck in the right wing by flak. He also made it back and landed safely in spite of damaged hydraulics.

Throughout September divebombing and strafing missions to the Halmaheras and Celebes were the rule. The 432nd Operational Diary refers to its September 19 mission to the Township of Amoerang in the Celebes as "a trigger-happy pilot's dream." Sixteen Clover Squadron P-38s took off and ranged over the target for forty-five minutes, setting fire to two warehouses as well as half the buildings in the town itself. Four or five barges were also strafed and smoke was seen up to 3,000 feet all along the coast.

Japanese Night Raids

At least six red alerts were sounded during the first half of September. The first came in the night hours of September 6 and early morning hours of September 7. One was heard at nine o'clock on the evening of the sixth and two more at 1:15 and three o'clock the next morning. Much noise was produced by the raids, but no damage was reported in the 475th area.

The next night three more alerts were enough to disturb the early morning sleep of the Satan's Angels, but the P-61 Black Widows of the 421st Nightfighter Squadron were able to claim two of the raiders, one of which was confirmed as a Ki-46 Dinah shot down.

On September 15 Morotai was invaded and the Japanese hold on areas to the south and east of the Philippines was further weakened. The Japanese Army 6th Air Division and 14th Air Brigade which had done much of the fighting in New

Working on the guns of a 432nd Squadron P-38 on Biak. . .

. . . and of a 433rd Squadron P-38.

433rd in flight over Pisang Islands, southwest of Vogelkop Peninsula, on the way to Amboina or Ceram. (Anderson)

433rd flight circling Boroka Strip on Biak. (Anderson)

Guinea were already deactivated. In October, the 4th Air Division which probably supplied the Dinahs and Sallys for the Biak air raids was sent back to Japan in preparation for the expected defense of the Philippines.

September Casualty

The only pilot to be lost during September was 432nd Squadron Lieutenant Walter Weisfus, the same one who was credited with capturing the Japanese soldier the month before.* On September 24 Weisfus was listed as being on a local flight in a borrowed 66th Service Squadron P-39. He took off at about 2:45 in the afternoon with enough fuel for about a hour and a half.

---

* NOTE: Five Japanese officers are listed in 475th records as POWs on Biak, the highest ranking being a Major Taruda. Perhaps one of these officers was the one captured by Weisfus and Griffith.

When he became overdue, Air and Sea rescue was notified and a search was begun. PBY and OA-10 Catalinas ranged over Biak and Owi throughout the next day, but Weisfus was officially declared missing in one of the odd quirks of combat flying; to be lost on a routine flight in the midst of a dangerous air war.

Boredom on Biak

Without the pressure of relentless combat missions the groundcrews eventually caught up on their maintenance duties and had more time to themselves. To use their time more productively the crews of the 431st would look for sources of fresh water which was in short supply. Well digging was frustrating because the sand became solid coral a few feet down. A few charges of explosives opened up the fresh water, but the liquid was milky white with dissolved lime and tasted as bad as the water from the lister bags.

One of the devices manufactured by the 431st Squadron Engineering Section was a diving bell designed for the shallow lagoons around Biak. The thing fit over the diver's head and chest and rested on his shoulders. Air was supplied by a pair of tanks secured by straps and a safety rope was thoughtfully included. The first to try the Rube Goldberg contraption came up after a few seconds gasping for air; the tanks leaked and emptied themselves almost immediately.

Left: On the way to Ceram. A 433rd P-38 heading out to divebomb a Japanese target, probably August or September 1944. (Anderson)

Lieutenant Robert "Pappy" Cline. He was one of the best pilots in the group and served well as flight leader and Ops officer. He became a captain by the end of October 1944 and was the nominal 431st commander sometime in December.

The men of the 475th were enchanted to find that the natives on Biak were accomplished flute players. Apparently the Dutch plantation owners had taught them how to play the instruments. There were bamboo sections in three sizes to play treble, alto and baritone parts.

Good sized crowds of Americans and natives gathered to hear native tunes, children's songs and even what seemed to be folk songs about the American visitors. The Yankee visitors were even delighted to hear Christmas carols in September.

New Aircraft

There were some changes in P-38 liveries by the end of September. The 431st had been using P-38Ls for some weeks in harness with the older J-20 model. Fifteen of the newer L models were received by the 432nd on September 29 to give the Clover Squadron a full complement. All the older J-20s

were being turned over to the 433rd which squadron had to wait until November for their Ls.

A diversity of opinion exists on the relative merits of the P-38L model and earlier versions of the Lightning. For the most part, P-38 veterans see little difference between the J and the L. Once a pilot became proficient in any sort of P-38, the fighter assumed something of a conditioned superiority over even single-engine aircraft.

European P-38 pilots were more enthusiastic about the high altitude capabilities of the new dive brakes and the enhanced ease of maneuver provided by the power-assisted aileron controls. Pacific veterans were already used to the heavy advantages that were inherent in the P-38: high level and climbing speeds, good diving characteristics at lower altitudes and firepower that was dreaded by nearly every Axis pilot.

The 475th added to the P-38's effectiveness by heavily disciplined tactics. It was possible for the Satan's Angels to drive into the enemy's midst and emerge with lopsided victories because of the formation integrity preached from the beginning of operations. Group records specify that the P-38L was a bit better in the climb and dive, but individual pilots were already well sold on the P-38.

Wrap-up

The end of September was just as quiet as the month could be. Maintenance, training and routine patrol and escort missions filled out the schedule. About twenty-five new pilots came into the group by the end of the month.

One of the enlisted men to leave for home at the end of September was TSgt Mel Allan. He flew down to Finschhafen to wait for the ship that was to take him home and stayed at the unpleasant base for nearly a month. During that time he was a magnet for every undesirable casual duty that came along; no matter where Allan tried to hide, the duty officer managed to find him.

Finally, after days of watching every ship that steamed into the harbor, Allan and about one-hundred other returnees boarded a Dutch steamer "(that) was little more than a rowboat", on October 28. The boat bobbed like a cork on the high seas, but Allan dined on Dutch lobster and caviar and he was on his way home.

October Losses

The month of October hardly opened when Possum Squadron lost one of its pilots to an operational accident. Lieutenant Charles Joseph of Greensburg, Pennsylvania was one of twelve 433rd Squadron P-38s preparing to takeoff for a divebombing mission of Fakfak when he was killed after his P-38 crashed during takeoff.

Another accident proved less serious for the 431st Squadron on the same day. In point of fact, while the incident was potentially quite dangerous at the time, the Hades crews came to regard it as one of their more humorous moments on Biak operations.

Lieutenant Wilson Ekdahl was a hardened veteran of the Satan's Angels who had yet to score his confirmed victory in the 431st victory column. The steady strain of combat flying from the first days of New Guinea operations with the 431st had marked this grim but determined P-38 pilot. He knew the law of averages was working against him, but flew operational missions time after time.

The 431st target for October 1 was Halong, Ceram. The mission was to glide bomb the place and Ekdahl had already tried to take off with a drop tank and 1,000 pound bomb when his engines began to act up. On a second try one of his engines cut out just as he was becoming airborne. He immediately cut the throttles and retracted his landing gear and waited for the impact with his deadly load.

Everyone in the area took cover while the P-38 skidded into the ground. It seemed an eternity before the bomb went off and scattered shrapnel in every direction. Later, when the fire in the wreckage had subsided a bit, Eckdahl's friends went looking with unbearable anxiety for his body. They didn't know whether to hug him or kill him again when he emerged from behind a piece of coral, wide eyed and wanting to know if it was safe.

The next day Clover Squadron lost one of its most aggressive and skilled pilots. Captain Billy Gresham took off

Jack Purdy was an ace and 433rd Ops Officer by the end of 1944. Here he poses bly the tail of Lewis's #170, probably soon after Lewis left the squadron and Campbell Wilson took over the squadron. (Purdy)

Joe Forster got his fourth aerial victory over Balikpapan, then set a record flying some 850 miles on a single engine. Ed Weaver does the same thing flying back from Haiphong a few months later. (Krane collection)

#187 getting fuel on Biak. (Anderson)

Jack Fisk's #191 on Biak. (Anderson)

P-38 #193 flown by Lt. Alan Sidnam of Kalamazoo, Michigan (Anderson)

from Mokmer strip at three o'clock in the afternoon on a local flight with only about enough gas for a three hour flight. After more than two hours had passed, squadron operations notified the Controller who tried to contact Gresham by radio without success.

Air and Sea Rescue was notified and a search plane was sent out. From the air a wrecked and smoking aircraft was sighted about ten miles northwest of Borokoe drome. The next day a search party was dispatched and Gresham's body was found in the gently undulating folds of his partially opened parachute that was still billowing slightly from an occasional weak breeze.

As early as October 5 the rumors about the coming invasion of the Philippines began to see tangible proof. Segments of the 475th began loading on the steamer S.S. Louis Weule and the 432nd camp, for one example, took perverse delight in tearing down the installations that had been so strenuously erected.

The destination, of course, was a strict secret, but the rumor mill strongly suggested that the target was the central Philippines. It would take a good bit of time to get all the different components of American airpower on the way to their destination, but the risk of exposure was part of the plan to surprise the Japanese in the Philippines.

Perhaps the most skillful maneuver General MacArthur made in his Southwest Pacific campaign was the initial landing in the Philippines. The selection of Leyte Island was both audacious and strategically advantageous. It was audacious because the Japanese expected a strike at Mindanao first; it was both within reach of Allied airfields north of the Vogelkop and required less time to reach by any sizable invasion fleet.

By October 6 the 431st was aboard another Liberty ship, the S.S. Charlotte Cushman, and joined the rest of the group waiting on the S.S. Louis Weule. Before the two ships would dock in Leyte the invasion would be on and the U.S. Navy would give the Imperial Japanese Navy what was perhaps its greatest defeat of the war in the Battle of Leyte Gulf. The Island of Leyte was also strategically positioned to cut Japanese forces in the Philippines off from each other if the Americans could hold the island against the expected determined resistance.

Balikpapan

One of the thorns in the side of Allied strategy that caused special anxiety was the supply of oil flowing into the Japanese empire from Balikpapan on Borneo. It was estimated that the Japanese had already shipped in many months supply of oil to the home islands to discourage any hope of an effective American Naval blockade.

At the beginning of the Pacific war the Japanese were widely believed to have no more than about a six month supply of oil. Even if these estimates are only approximately accurate it is apparent that the oil fields of the Pacific islands were enriching the Japanese reserves. Thus, the necessity of dealing with the matter by strategic bombing was sorely felt.

One unescorted mission from New Guinea in October 1943 failed to achieve any appreciable results. A year later the gains in Pacific territory and the contribution of Charles Lindbergh to the range of escorting fighters had made more concentrated attacks on Balikpapan possible.

The first B-24 strike escorted by 35th Fighter Group P-47s and 49th Fighter Group P-38s was mounted on October 10, 1944. Over 100 Fifth and Thirteenth Air Force B-24s struck at the refineries, losing four of their number in the generally successful operation. The P-47s claimed twelve and the P-38s claimed six more, raising some envy among the aerial combat-starved fighter pilots who were left behind on Biak and Morotai.

There were enough 432nd Squadron pilots left on Mokmer strip to make up a sixteen plane flight for the last B-24 escort to Balikpapan on October 14. Clover P-38s would escort the Thirteenth Air Force B-24s coming in behind the Fifth Air Force bombers covered by the P-38s of the 9th Fighter Squadron.

It was a splendid opportunity from the point of view of the Clover pilots who would be able to fly the mission. Meryl Smith came down from Headquarters to lead the escort as Clover Red One. The 9th Fighter Squadron escort was led by Major Wallace Jordan and some of the most distinguished aces of the area.

Two of those aces were flying together in Captive White Flight; Majors Jerry Johnson and Tom McGuire. Somehow, McGuire had managed to get invited along on the mission with the 9th Fighter Squadron since his own squadron was almost entirely packed away on the S. S. Charlotte Cushman. He was happy enough to use a borrowed brand-new P-38L from his host unit and looked forward to advancing his personal score that now stood at twenty-one aerial victories.

The 432nd took off from Biak on the afternoon of October 13 to land at the Morotai staging point. At 6:45 the next morning thirty-three P-38s of the two squadrons took off on a round trip of more than a thousand miles. The Lindbergh effect would have its greatest test to date.

Weather en route to the target was almost absolutely clear. When the 9th Squadron reached Balikpapan at 10:30 the clouds were heavier, and the white four to six-tenths masses seemed to offer any intercepting Japanese a reasonably good line of concealment.

Captive White Flight – with Johnson as Captive White One and Tom McGuire as White Two – was at 19,000 feet

about fifteen to twenty miles from the target when Japanese fighters were seen. About twenty very dark-camouflaged Tojos and Oscars were outlined against the sky above, rolling and looping as if to get a better view of their American objectives below.

Johnson led McGuire up into a left climbing turn to get an advantageous position. McGuire fired at one Oscar that got away in a fast turn to the right. With his fighting blood up, however, McGuire looked around for any other likely prospects.

One Oscar seemed to be milling around to no apparent purpose and Johnson seized on the opportunity. He drove right in the tail of his target and fired one telling burst, but had too much forward speed and overran the Japanese. McGuire didn't need any more invitation to throttle back, get behind the enemy fighter and shot it up until there was fire erupting from the fuselage and the pilot bailed out.

The Japanese pilot's parachute opened after he had fallen just behind the bomber formation. Below was the airfield at Manggar, about ten miles from Balikpapan and probably the base of the pilot now dangling in his chute. Jerry Johnson saw McGuire's kill and then proceeded to shoot down another Oscar from dead astern.

More interceptors were coming in now, and the two P-38 aces had all the targets they could want for the moment. Two Tojos came into the battle area and one of them made the mistake of turning toward the bombers and ended up another victory for Johnson. McGuire made a twenty degree deflection shot at another Oscar and followed it down to about 15,000 feet where he last saw it spinning and smoking. This Japanese fighter went into the records as a probable.

Johnson then followed McGuire into a left turn onto the tail of what they later decided was a Hamp. A heavy concentration of bullets and shells from McGuire's guns tore pieces off the enemy's tail and both P-38 pilots saw the plane burst into flames.

At that moment McGuire and Johnson became separated, but McGuire was not about to let opportunity slip from his grasp. He fired at any target that presented itself, usually with a high-deflection shot before breaking away from some attacking interceptor.

McGuire looked around and saw one Tojo being shot down by a P-38. This kill was probably the one made by Lieutenant Edwin Cooper of the 9th Squadron, since he is the only other pilot of a P-38 credited with a Tojo on this mission. At that time Cooper latched onto the wing of McGuire's P-38 as it passed and the two went into the fray.

The two P-38s quickly got onto the tail of another Tojo at 14,000 feet and drove it down to the cloud level at about 3,000

Billy Gresham was a hot and aggressive pilot who could have gone on to further distinguish himself – or perhaps just go home within a few weeks, if he wasn't lost on a routine test hop, October 2, 1944. (Anderson)

feet. McGuire opened fire from 400 to 200 yards and saw smoke begin to stream behind the black/green fuselage of his target. Apparently, the fast-diving Tojo was estimated to be moving at about 500 miles an hour, giving McGuire quite a time before he was able to close the gap.

Just as the Tojo was slipping into a cloud McGuire fired again and drew more dense, black smoke. Not to be deterred, the determined P-38 pilot followed right behind and emerged at the bottom still on the tail of the Tojo at about 1500 feet. Another burst of gunfire and the Tojo began to burn. McGuire turned to avoid running into his target, and watched the silver belly of the now inverted Tojo fall away into the water. Cooper had watched McGuire's twenty-fifth kill crash into the sea.

Clover Squadron had been listening to the opening engagements of the 9th Squadron with the Japanese interceptors over the radio. The sixteen P-38s had passed the Thirteenth Air Force B-24s about forty-five minutes before reaching the target and were between them and the battle going on ahead.

Colonel Smith was about ten minutes from the target when he could hear another dogfight starting and saw the puffs of antiaircraft fire over the target itself. By the time he and his Clover Squadron P-38s arrived over Balikpapan the bombers were already turning away from the bomb run. Flak was still bursting in the air and the powdery octopus shape of white phosphorus bomb clouds were also evident.

McGuire and the other P-38s of the 9th Squadron were just in the process of dispersing when Clover Squadron arrived, but there were still at least six Japanese fighters identified as Zeros and Hamps milling around at about seven or eight thousand feet "between the airfield and the town and slightly out to sea."

Clover Red Flight immediately dived down to the attack with Clover White and Green coming in trail. Clover Blue stayed above as top cover.

Smith was diving on the enemy fighters, now. He could see them in their brand-new camouflage paint, alert to the attacking P-38s and ducking into the clouds. Clover Red Flight couldn't get any decent shots on the initial bounce.

The third pass that Smith made was at the Hamp that carelessly dove right in front of Red Flight. With a well-aimed burst, Smith got the bottom of the Hamp's engine cowling to emit bright flames. Captain Richard Kiick was flying Clover Red Three and saw the Hamp burst into flames. Lieutenant Dallas Peavyhouse was bringing up the rear of the flight and also saw the Hamp go down.

Joe Forster had moved from element leader in Green Flight to White Flight because of a SNAFU early in the mission. He had watched Red Flight taking on the elusive Japanese below before his own flight got into range. The entire flight made a coordinated attack or two before Forster took his wingman, Lieutenant Dean Olson, on their own unsuccessful bounces.

While he and Olson were in a turn Forster noticed a Zero coming down on a Clover White Flight P-38's tail. Immediately, Forster took a long shot that discouraged the Zero, but no other reactions indicated that the Japanese was hit. Another Zero crossed the nose of Forster's P-38 and all he had to do was push over and fire a one second burst that he and Olson saw was enough to explode the fighter.

By eleven o'clock the battle was dispersing, because the bombers were already well on their way home and the P-38 pilots were beginning to sweat out their fuel. McGuire and Cooper had started back at ten minutes before eleven and would land at Morotai three and one-half hours later. Clover Squadron left at precisely eleven o'clock and was back on Morotai by two-thirty.

433rd lineup in the last few months of 1944. #179 is the first in line and #192 is next.

Clover Red Flight chased off Japanese interceptors who were halfheartedly attacking the bombers while they were leaving the target area. Meryl Smith saw one P-47 that had been inadvertently damaged by a P-38 crash en route to Morotai. Fortunately, the P-47 pilot managed to bail out and get into his rubber dinghy in the water about eight miles west of Karong Point in the Celebes. Smith called the Catalina rescue service codenamed Daylight 12 and circled the downed pilot until the Catalina found him.

Joe Forster's Long Ride

One of the Japanese interceptors that Clover Red Flight had chased off the tails of the bombers was aggressive enough to attack Colonel Smith. Joe Forster saw the threat to his leader and moved to the defense. He was astonished when the Zero pulled up in a tight loop and turned sharply onto the tail of his own P-38. The enemy plane came quickly into firing position from behind and slightly to the left. Before Forster could do much about it the guns of the deadly Japanese fighter opened up.

There was an explosion in Forster's left engine. Almost by instinct he rolled the P-38 to the right and went straight down. When he looked around for the Zero it was gone; gratefully he thanked the great diving speed of the P-38 that had once again saved its pilot.

However, Forster was not yet out of the woods. His left engine was spitting oil from a broken line and the pressure was going down. There was nothing left to do but feather the engine and face a long ride home on single engine. If this moment seemed daunting to Forster he would later realize that no single-engine fighter would have any chance in the same situation.

Forster has confidence in the ability of his P-38 to make the flight and he has at least one other Clover fighter to escort him as a check on navigation and deterrent to any unlikely Japanese aircraft that could happen along. As it turned out, Forster landed back at Morotai about an hour after most of Clover Squadron. He was credited at the time with something of a record 850 miles on single engine in a P-38.

Meanwhile, the sea echelon of the 475th sailed from Biak on October 8, heading for Wakde Island. On the first night out many of the crews abandoned the hot and smelly inside of the ships for the relative comfort of the open decks. A rainstorm caused a deluge that disturbed the sleep of the unfortunate travelers and sent them scurrying for shelter.

Around noon the Louis Weule pulled into Maffin Bay to pick up a cargo of 1,000 pound bombs. It took four days including the day the air echelon flew the Balikpapan mission to load the bombs via LCTs from shore. The last bomb was aboard around four o'clock in afternoon on October 15 and the ships were underway almost before the loading crews could clear the decks.

A few hours later the 475th was in Humboldt Bay and the crews were awestruck at the number of ships assembled for the Philippine invasion. October 16 was a day of calm for the men on the ships while the great invasion fleet went about the business of loading and assembling. Early on the morning of the eighteenth the fleet once again went to sea.

The 475th crews could see ships all around them when their place in the convoy was reached. It seemed that the 475th

Andy Anderson shows the approved method of mounting a P-38: First, the left foot on the bottom step, then the right foot on the next step and up on the left side of the gondola with a smart left swing. (Anderson)

Anderson ready to start up on Biak. (Anderson)

ships were in the center of the pack. Colonel Smith had flown from Biak to see that six enlisted men of the group were not part of this impressive display; he handed the lucky batch of men their rotation papers and they set off as a group in the direction of home.

On October 19 the convoy crossed the equator, heading north in the early predawn hours. The venerable custom of introducing pollywogs into the court of King Neptune was administered to the men who had never, or at least could not prove that they had, crossed the sanctified line. Those men were required to run the length of the ship's decks with their hair clipped and their bodies painted the most outlandish colors.

The next day Captain Reginald Hawley, the 475th Intelligence Officer, announced over the public address system that the convoy would be landing at Leyte Island within the next few days. First news of the Philippine invasion came from the Japanese broadcasts. The enemy version had ten American aircraft carriers sunk, four battleships and many cruisers and destroyers also claimed and 1,000 American aircraft shot down. The landing forces had been thrown back into the sea and were fighting knee-deep in the water.

It would take some time for the men on the convoy ships to learn anything like the real situation. Whatever the facts were the invasion of the Philippines was on and the Satan's Angels would be part of it within a few days.

#198, "The Dorothy B" on Biak. (Anderson)

Chapter XIII

CLIMAX IN THE PHILIPPINES NOVEMBER-DECEMBER 1944

Colonel MacDonald returned to the 475th Fighter Group on October 13, 1944. Although he has no recollection of it, group legend has it that he gave McGuire a thorough chewing out for going off on a mission with another fighter group while his squadron was preparing for a major move – not to mention the fact that three air victories scored by McGuire were added to the 9th Squadron tally.

By this point in the war McGuire was consumed by the urge to become the top American fighter ace. Dick Bong was back in combat, albeit surreptitiously aside from instructor duty, and was officially credited with thirty victories. McGuire had twenty-four confirmations, but needed every break to overcome Bong's easy facility with the P-38 in battle.

The news aboard the ships of the convoy was encouraging. The first invasion troops had established two wide beachheads on Leyte's eastern shore and the First Cavalry and 24th Infantry Divisions held Tacloban airstrip and the high ground on the northern coast.

On October 24, when the Battle of Leyte Gulf was in its full fury, the 475th men were subjected to two Japanese air attacks. The first occurred just as the ships were pulling into Tacloban to put reinforcements on the shore. Tracers arced overhead and great geysers of white water erupted from exploding bombs. It was nerve-wracking for the men on the ships to realize that they were sitting on many tons of high explosives and gasoline.

Almost before the first raid was over, another group of Japanese planes came and gave the convoy another pounding. One of the first of the terrible examples of the new kamikaze tactic was experienced when a Japanese bomber sank an LCI by crashing into it and exploding like an orange fireball.

When the air became silent once again the two ships bearing the 475th water echelon made for their destination at Violet Beach and anchored off the town of Dulag. Those aboard the ship spent the night on them and awoke to yet another Japanese air raid. The 432nd Flight Surgeon, Adjutant and one pilot, Lieutenant Edgar Williams were seriously wounded and had to be evacuated.

Two more harrowing air raids were experienced during the day, but no additional casualties were suffered. The Navy came up to the Louis Weule in one of their Lighters and took off more than one-hundred groundcrew to service their aircraft that at the moment were the only air superiority on Leyte itself.

Tom McGuire Scores Over the Philippines

It was a departure from usual American planning to seize an objective beyond effective land based airpower. This time the Japanese were caught flatfooted and the initial thrust into the Philippines

Above and opposite: Champlin (112) Tilley (122) DuMontier (125) and Pietz (126 as yet unnumbered) over Leyte at the end of 1944.

was more than successful. By October 30 the entire northeast corner of Leyte was nominally in American hands from Leyte Valley to the hills overlooking the route to the main Japanese base at Ormoc on the western shore.

To relieve the burden of fighting without air cover, the 49th Fighter Group was ordered up to Tacloban on October 27. Dick Bong went up with his old 49th Group and quickly raised his score to thirty-three confirmed by October 28. Many of the soldiers on the ground swear to this day that they could pick Bong's P-38 out of the swarms of P-38s flying overhead. At any rate, every American on the ground was grateful to see land based U.S. warplanes overhead.

It was inevitable that Tom McGuire would get into action, and he found a way on November 1. The 431st Squadron was required to ferry in seventeen 49th Fighter Group P-38s and McGuire took his rightful place as leader of the formation. He took off from Buri Strip on Morotai at 7:30 in the morning with sixteen other Hades pilots behind him in their P-38s, all heading for Tacloban.

Arriving over Tacloban at ten thirty through perfectly clear weather McGuire was told by the controller that a red

alert was on. It was good news to the eager P-38 ace and he answered in the enthusiastic affirmative when the controller further asked if the P-38s had enough gas to patrol the area.

The P-38s took up patrol ten miles southwest of Tacloban at 10,000 feet. McGuire sent up his Hades Blue and Green Flights to 15,000 feet as top cover.

At fifteen minutes past eleven a single Tojo appeared about a thousand feet above McGuire's two flights at 10,000 feet. The Japanese fighter was approaching from the right rear of the P-38s and apparently sensed that he was sandwiched between two powerful enemy formations because he suddenly turned around and headed for cloud cover.

It was not much use trying to outrun the determined McGuire who caught up with the Japanese just south of San Pablo. The first deflection shot from three hundred yards back caught the Tojo in the cockpit with a 20 mm shell. Two more bursts were fired from one-hundred to fifty feet and the Tojo's tail began to disintegrate. McGuire and another pilot in his flight, F/O Edward O'Neil, both witnessed the Tojo spinning and crashing into the hills below.

Number 192 in November on Biak. (Anderson)

A 433rd Squadron P-38 during the last days on Biak. (Anderson)

McGuire was still resolute and led his P-38s on patrol until most of them ran short of gas. The P-38s landed at Tacloban at noon with McGuire exultant over his latest victory. Again, group legend suggests that he was disappointed only in hearing that Bong now had thirty-three victories to his twenty-five. For the next few weeks it would seem that McGuire had to pace eight victories behind his rival.

Whirlwind War

If McGuire had to play catch up to Bong at least he was in a hot war once again. The Japanese were desperate to regain the momentum in the Philippines they realized the ease with which the islands could fall to a well-directed attack and sent waves of aircraft in frantic raids.

Most of the pilots who flew these raids were young and inexperienced Japanese who replaced the redoubtable veterans lost to American Army, Navy and Marine pilots who were now closing in on the Philippines. The result was a killing ground for the Americans who claimed many hundreds of the enemy during November and December 1944.

The crewchiefs and other groundcrew who went up to service the Navy planes that were forced down on Tacloban were back in the 475th camps by the end of October. Those 432nd Squadron crews were just in time to turn around and go back to Tacloban to service their own P-38s that had landed there on November 2.

When the 432nd landed at Tacloban around 12:30 in the afternoon it found itself in the middle of the wild action that was just beginning to build. Joe Forster was in the middle of stretching his legs after the long flight from Morotai when the red alert was sounded. He was immediately scrambled as leader of Green Flight without benefit of even knowing the area where he was supposed to find the enemy.

On his wing was Lieutenant Walter Freeman as Clover Green Two. Freeman was a youngster who had been in the 432nd Squadron only a few weeks, but he was in his first action. Forster led him up to 12,000 feet and tried to get a vector from the Controller. Unfortunately, the map in Forster's cockpit was not coordinated with the Controller's grid, so the P-38 flight leader turned out to sea where the greatest activity seemed to be centered over the Japanese convoy at Ormoc Bay. Forster watched a group of B-24s hammer the transports

Andy Anderson finds a new mode of transportation while his P-38 is being tended to. (Anderson)

Sgt Ruiz and the man he trusts with his airplane – on official occasions only. (Anderson)

*433rd pilot Bob Morrell sitting in the cockpit of VIRGINIA MARIE.
(Anderson)*

Traffic sign at Dulag, Leyte.

and the landing beaches before he got word from the Controller to fly a heading along the northern tip Ormoc Bay. After a bit of patrolling along the area Forster saw a beautifully green-camouflaged Oscar at the top of its zoom out in the front and to the right. Apparently the Oscar did not see Clover Green

Flight, but was turning in the right direction to get onto the tail of Clover Green Four. Forster used his dive brakes to get into a tight enough turn to head off the Japanese fighter. The Oscar pilot had seen the number four P-38 and reefed the nimble aircraft in a tight turn to get a shot.

Grady Laseter sometime earlier in the war. He must have experienced some disorientation that caused him to reverse course and collide with Dahl on November 10, 1944.

Unidentified pilot of the 432nd with a monkey clinging to his neck. After Dahl walked back into camp with his pet monkey it was all the rage to have the little animals for pets and mascots. (Krane files)

Fred Champlin with a monkey he associated with long before Dahl made it popular. Champlin claimed that he taught the little critter how to fly and shoot and that it was responsible for his victories! (Krane Files)

Before the Japanese could open fire, Forster had completed his turn and was closing on the Oscar's tail. The Japanese split-S'ed and Forster got a lesson in camouflage when the Oscar blended in perfectly with the jungle below. Following the logical path of the Oscar. Forster again sighted it at about 7,000 feet. Pushing the P-38 to its limits he caught the running target at 3,000 feet south of Ponson Island and shot it down into the water with two short bursts. It was Forster's fifth victory.

The tempo of the war in the Philippines refused to flag during the first days of November. in addition, the weather seemed to augment the fury of the Japanese attacks with rainstorms that produced thick, clinging mud. The 432nd Squadron began operating from Tacloban and Bayug strips while the facilities at Dulag were being prepared in spite of opposition from the elements as well as the enemy.

Throughout the first week of November the 432nd flew patrols over the shipping in Leyte Gulf from dawn to dusk. A typical day was like November 6, in which twenty-four sorties were flown by Clover Squadron. Operations began with four P-38s off at six o'clock in the morning and ended with the last four landing by five-thirty in the afternoon at Bayug.

Japanese air raids seemed to decrease after the P-38s arrived on Leyte, causing the Americans to be lulled into a sense of false security. At six in the morning of November 7 a Japanese raid came over Bayug and literally caught the Yanks with their pants down. The Japanese fighters that came over were mistaken for P-40s and no antiaircraft batteries responded to the attack.

Several clusters of the infamous "daisy-cutter" bombs antipersonnel explosives that shot fragments of steel in every direction at a low height damaged three of the 432nd P-38s.

The camp at Dulag was also treated to a heavy bomb dropped by another early morning visitor in the nearby vicinity.

Nevertheless, the patrols continued through the weather and interference by Japanese air raids. Another twenty-four sorties were flown on November 7 and six more on the next day in spite of a solid weather front that towered from the deck to as far as anyone could guess.

The poor weather was no obstacle to Joe Forster claiming his sixth aerial kill. He took off at a few minutes after six in the morning with Lieutenant Lakin Parlett on his wing. The mission was to cover an LCI off Patcan Island, but the two P-38s barely had time to turn after taking off when three bursts from a 40 mm antiaircraft gun were seen over Dulag.

Forster watched the bursts of flak intersect and plotted the course of the enemy aircraft that had now thoroughly aroused the ground defenses. The Japanese fighter was now visible flying due west and approaching San Pablo strip number 2. Forster called for the antiaircraft to stop, but had to wait until

Jim Barnes by his BARNES STORMER.

attacking P-38 continued to drive closer behind the Oscar. The Japanese pilot bailed out.

Fairly accurate fire was coming up from the Japanese destroyers below, and with his fuel running low MacDonald decided to get the P-38s out of the area. He turned the formation over to Perry Dahl and returned to base with his wingman.

Dahl subsequently chased a Zero into the clouds before the Clover formation ran into a gaggle of about twelve Tonys breaking in and out of the clouds. The weather was beginning to clear over that portion of Ormoc Bay and Dahl ordered his P-38s to increase power and climb into the area of better visibility.

Before the clouds began to disperse Dahl had been anxious about the vulnerability of his flights, but now he was able to take the advantage. The P-38s circled in the open sky and waited. within a few minutes the Tonys emerged from the

Above and below: Lt. Tom Oxford

the Japanese had flown clear of the flak bursts before he dived down to the attack from 2,000 feet, drawing war emergency power as he did.

Identifiable as an Oscar when Forster drew within range over the top of the Catmon mountains, the Japanese fighter began taking hits just before it slipped into the clouds. Forster followed the enemy in and out of the clouds, getting deflection shots until he was sure the Oscar was badly hit. Forster pulled off to the side of the cloudbank and both he and Parlett saw the Oscar roll out of the murk with flames trailing behind and the pilot bail out.

The biggest fight encountered to that date in the Philippines by Clover Squadron came on November 10. A scramble was ordered and twelve Clover P-38s responded by taking off at 0815. There were plots coming in that detected enemy airplanes over Ormoc Bay.

Colonel MacDonald was leading the scramble with Lt. Calvin Anderson as his wingman. The 475th commander took his P-38s to 10,000 feet and arrived over the enemy shipping in the bay a few minutes after nine o'clock. One Oscar was visible fading in and out of the clouds and rain, followed by three other unidentified aircraft.

Without much hesitation MacDonald chased the Oscar until it disappeared again into the clouds. Getting out of the cloud himself and gathering the P-38s together again, MacDonald noticed the Oscar once again emerging into the open. Again acting quickly MacDonald closed to about two hundred yards behind and fired a 30 degree deflection shot. A few more bursts and the Oscar began to burn. When the

Louis DuMontier

Jim Moering

muddled clouds and presented themselves as gleaming silver with brown and green daubed camouflage targets.

The P-38s came down from the direction of the sun and certainly surprised their prey. Dahl selected the leader and fired a 45 degree deflection shot from above. This Tony took hits in the engine and cockpit before it fell out of control. Most of the P-38 formation could see it crash below.

Above and two below: #114, Tom "Pepper" Martin

That is, those P-38s who weren't busy shooting down other Tonys. Lt. Charles "Rat" Ratajski was leading Clover Blue Flight in the bounce and thought he could see five Tonys going down in flames before he could get his sights on one of the enemy.

When he did center his aim he found that he didn't have much trouble climbing with the Tonys at about 175 mph. However, he missed both Tonys he shot at in the climb because they managed to roll over and escape beneath him.

Another Tony was observed by Ratajski just as it overshot Lt. Don Willis's P-38. Ratajski was able to confirm the Tony destroyed when Willis got in good strikes that tore the left wing off the Japanese fighter. Then Ratajski saw another Tony ahead and a little above. Making a fast approach on the Tony, Ratajski drew within range and set it afire with a good three-second burst. Other members of Blue Flight saw the Tony crash into the water.

Ratajski also witnessed Clover Red Three Lt. Dean Olson shoot down another Tony in flames. Willis had to feather one of his engines and Ratajski escorted him out of the area, but

Two above: #116, A.R. Neal

Ratajski also managed to see four parachutes descending and two more aircraft explode in midair.

Lucky Dahl

One of those parachutes could have been saving the life of Perry Dahl. Ratajski had also seen two P-38s collide, but couldn't be sure which of his comrades he had just lost. He had to wait until the squadron landed again around twenty minutes after nine at Tacloban before he knew who was lost.

The 432nd Squadron operational diary had a somber entry regarding the midair collision: "We didn't come out of the fracas unscathed, however, as two (2) of our planes, flown by Lts. Dahl and Laseter, collided in midair while engaging the enemy. Their loss was a hard blow for the squadron . . . (although) there was some slight hope for one of them . . . as a parachute was seen to open beneath the wreckage of the planes."

That faint optimistic note at the end was eventually realized. Perry Dahl had once again emerged from a deadly situation and prevailed over the law of averages.

#117, Floyd Fenton

#115, Hal "Pete" Madison

#118, Bob "Red" Herman

Two above: #120, John O'Rourke

After Dahl had shot down his Tony he broke to the left with Don Willis close behind him. Somehow, Grady Laseter had turned around in the formation either because of mechanical trouble or hits from enemy fire – and came straight for Dahl's P-38. At first Dahl thought it was a Japanese "self-blaster", but later learned that Laseter had slammed into his P-38 and cut off the left wing and tail booms.

The P-38 was on fire, giving Dahl some bad burns before he could get out of the cockpit and open his parachute. He landed in the water near four of the Japanese destroyers. Willis

#121, Pilot Officer O'Neill

#119, Lt. Woodruff

#124, Crewchief Richard Vander Geest

was by this time sitting in a "flying sieve" because of battle damage and was as much in need of help as Dahl; there would be no help from that quarter.

One of the destroyers opened fire with a light machine gun on the helpless American after he landed in the water. One

bullet grazed his head while he opened his life raft and Dahl decided to play dead, hanging by one arm over the side of the raft. The destroyer steamed close by and Japanese sailors examined the seemingly lifeless body. Dahl clung to the raft with his face in the water and could hear the thrum of the idling

#123, Douglas S. Thropp

#124 Pilot: Lt. Jack "Foxy" Olson

By dusk the wind had blown him in toward shore. He saw canoes coming for him, but didn't want to risk being picked up by the Japanese. Making it onto the beach by himself he got into the trees and himself until he could assess the situation. He had been in the water for nine hours and didn't want to make a mistake as soon as he reached land.

Fortunately, he was quickly picked up by filipino civilians who took him to a house where they hid him, fed him and

Above and below: #125 MADU IV, Louis DuMontier

engines mixed with the voices of curious Japanese crewmen. He must have looked dead or at least mortally wounded with his neck and shoulder burned black. He dared not breathe to give the impression that any life was left in him.

Suddenly, the destroyer's engines raised in tempo and the loud drumming of antiaircraft guns opened up. The Japanese warship was hustling off to Dahl's great relief. He dared to look up after the menace had left and silently cheered to realize that some B-25s had arrived to give the Japanese hell, just as they were giving it to him moments before.

When both Destroyer and its B-25 attackers had left, the silence was broken again by the sound of a single aircraft engine. Soon a lone Tony began circling the rubber raft like some airborne shark. The fighter made one strafing run, but perhaps was low on ammunition or didn't want to get caught low and slow over the water because he then flew away to the west.

Sometime later the B-25s finished their attacks and the destroyers returned. Dahl assumed the same:position as before, but this time nobody evened slowed down to take notice of him. He watched them vacate the area with valiant discretion.

#125 MADU IV, detail

#136, George W. Skipper. Man in photo is probably crewchief.

tended to his wounds. He slept soundly after the grueling experiences of the day were set aside and the danger of intruding Japanese was not immediate.

The next morning he trekked to Santa Cruz, then went by boat to Matlang. From there. he was transported to the 3rd Battalion of the 96th Infantry Regiment Aid Station. Lieutenant Juan Rosete, the station surgeon, took care of his wounds until December 3, when several other shot down Americans, including 433rd Squadron pilot Lt. LeRoy Ross who was downed on November 24, started out for Cabaliwan, about ten miles west of Ormoc Valley.

Lt. Rosete had made contact with a PBY on December 7 and arranged to have Dahl and the others picked up at Cabaliwan on the tenth. They traveled back the way they came and rendezvoused with the PBY at Matlang on the morning of the tenth. The exhausted bunch of survivors were all flown back to Tacloban.

Dahl eagerly accepted a ride in the back seat of an L-5 and was flown back to Dulag. The Clover Squadron boys were

pleasantly surprised when their aptly nicknamed "Lucky" Dahl came walking leisurely into camp with a monkey on his shoulder.

Possum Action in the Philippines

LeRoy Ross had been in one of the first 433rd Squadron engagements after it flew to Leyte during the second week of November. He had shot down a Tony on November 24 and was subsequently shot down himself on a later patrol.

Possum Squadron began operations without incident over Leyte Gulf on November 14. Two flights of four patrolled the Gulf without finding any opposition. Seven missions were flown on November 16 with nothing more than a single Tojo being engaged.

However, four of eight Possum flights ran into enemy aircraft on November 19. At about 7:15 in the morning a single Jack (Mitsubishi J2m Shinden) was claimed damaged and a

#128, J.W. Rohrer

#137, Howard Max

#138, R.G. Haley

Lily and Oscar were also damaged at 3:30 in the afternoon. First blood over the Philippines for the squadron was claimed by Lt. Pierre Schoener when he shot down a Val south of Dulag later in the afternoon.

Cal Wire led the last patrol of the day with Lieutenant Gallagher on his wing. The four Possum Squadron P-38s took off at ten minutes to four in the afternoon to patrol the west coast of Leyte. They were scheduled to be over the area at 10,000 feet from four o'clock until about five-thirty.

The Fighter Controller directed Wire to a fight that was supposed to be going on, but the P-38s found another Japanese formation heading due south. There were eight dirty-brown-mottled Oscars, six dead ahead at the P-38 altitude and two others below and behind the gaggle of six. Apparently, the two Oscars sighted the P-38s and went into a fist dive below the higher batch of Japanese.

Wire calculated that the lower two fighters were decoys for their comrades above. One of the higher-flying Oscars began waggling his wings to signal the others and the whole formation began climbing, the outer fighters breaking gently to the right and left. The right Oscar became Wire's target and he opened fire at two hundred yards. When the Oscar made a split-S and tried to dive away Wire was pleased to find that his new P-38L with aileron boost and dive flaps was able to stay with the nimble little fighter and even drew within 100 yards. A two-second burst from the P-38's guns sent the Oscar into the water at full speed.

#138, "Miss Skeets"

#139. Lt. Werner was one of the last of the combat 431st pilots and took the last numbered P-38

433rd Squadron records claim that the P-38L on this mission was able to leave the P-38J-20 behind and was about 6,000 feet ahead in the dive. Wire doubts that a great difference existed between the J and L model and attributes the spacing of leader and wingman to normal difficulties of the wingman trying to keep up with his leader. "Both (the J and the L) were beautiful", he stated in a recent letter to the author, "the best fighter planes in the world."

Subsequently during the mission, Wire slipped in behind another lone Oscar and closed to within fifty yards. The P-38's 20mm fire smashed the engine and cockpit, the Oscar crashing and exploding into the ground south of Baybay.

The Japanese were operating a new naval fighter in the Philippines that the 433rd Squadron would encounter within

SSgt George J. Haas on the tail of his P-38, BARNES STORMER.

the next few days. Japanese Air Group 341, operating the Nakajima N1K Shiden, codenamed "George" by the Allies, began sweeps around the Leyte area in November. Commander Aya-o Shirane was leader of the 701st Hikotai at Mabalacat, near Clark Field, Shirane's pilots were totally inexperienced although he had been in action almost from the beginning of the Pacific war. He was a popular leader who inspired confidence in his green troops. By the end of October his unit would be ready to engage the Americans and Possum Squadron would be the immediate opponent.

November 24 was a day of wild and confused action. Two P-38s of the 433rd, the one flown by Ross and another flown by Mort Ryerson that crashlanded two miles west of Dulag strip, were lost during the day. In return, the squadron claimed four Jacks, two Tonys and a Zero.

Close examination of the day's actions has convinced Japanese historians that the 433rd was, in fact, engaged by Shirane's 701st Hikotai. The rapid nature of the various combats may have led to confusion between the generally similar George and Jack. One of the 433rd reports suggests that the Jacks could have been flying with Tojo fighters, a highly unlikely circumstance.

At any rate, the highly regarded Shirane was shot down during the day's action and killed in action. He could have been claimed by either Ross or Ryerson who in turn were shot down themselves, both surviving with little injury.

Ace Race

On November 10, both Bong and McGuire each shot down an Oscar. For McGuire it finally meant that he went into the select group of American fighter aces who claimed twenty-six or more aerial victories. For Bong it was an added margin to insure that he would always be the American ranking ace.

Maintenence on 433rd P-38 #184 BOBBSEY. Note the unusual anti-glare panel. (Anderson)

By the first week of December Bong had moved from his old 49th Fighter Group to the 475th Group. General Kenney always claimed that he kept the two aces together because they wanted to fly together and the public relations aspect was priceless. In truth, McGuire was frustrated by Bong's consistent hold on the status of ranking ace and provoked him mercilessly. At some point the usual calm exterior presented by Bong was fractured and he summarily moved out of the quarters that he shared with McGuire.

December 7, 1944 was the day that an American landing was made at Ormoc Bay to break the stalemate caused by stubborn Japanese resistance on Leyte. Particularly savage air battles developed throughout the day and the fighters of the Fifth Air Force claimed over fifty Japanese aircraft for the loss of a single pilot.

McGuire had scored earlier in the day when he and Major Jack Rittmayer went out on a two-man sweep as Daddy Green

Flight. (The 431st had changed its call sign from Hades to Daddy by this time.) With twenty-nine confirmed victories to his credit, McGuire was trailing Bong's thirty-six by seven.

A little after two o'clock in the afternoon Daddy Green Flight was off again, this time with Dick Bong as Daddy Green Three and Lt. Floyd Fulkerson as his wingman, Daddy Green Four. The mission was a convoy patrol south of Ormoc, but the weather was so bad that the four P-38s were stationed northwest of the convoy at 4,000 feet.

Bong was the first to call out the enemy when he sighted a Sally bomber at five minutes after three. Once again McGuire

Champlin, Hart, Pietz and Monk spend a moment in the Philippines with Champlin's monkey.

Joe Forster, Dick Kimball and Larry LeBaron. Forster and Kimball scored during the big battles of December 7, 1944 and Le Baron got three in one day in March 1945. (475th History)

McGuire being strapped in sometime in December 1944. The notorious battered cap is being held on the head of the crewman for the moment; it had better have been nearby if the wrath of McGuire was to be avoided.

was frustrated in trying to locate the enemy, so Bong gave chase and shot the Sally down on the northeastern tip of Bohol Island. Once again Bong was ahead by eight with the destruction of his thirty-seventh Japanese aircraft.

McGuire resumed the patrol with an urgency borne of his desire to overtake Bong's score. About half an hour later he attacked a Japanese divebomber and pressed the attack until he was driven off by flak from friendly ships below. Jack Rittmayer was able to get behind the divebomber and shoot it down.

About thirty minutes after that, McGuire sighted five Tojos coming in at 2,500 feet from the north, just under the overcast. McGuire again went through friendly antiaircraft fire, but this time he pressed on and shot at the left side Tojo until it crashed just short of the convoy.

Another Tojo was shot down by Fulkerson who watched his victim explode violently over a formation of LSTs. McGuire and Fulkerson witnessed Bong make a head on pass at another Tojo that was mortally hit and crashed into the water.

Now it was thirty victories for McGuire, but Bong still led with thirty eight. McGuire reformed the flight and set out on patrol once again. The controller vectored Daddy Green Flight over a damaged destroyer where a single Oscar was observed at about 6.000 feet. Daddy Green gave chase to get the lone Japanese fighter, but Daddy White Flight came into the action and stole the victory away.

Fred Champlin was leading Daddy White with Lt. Tom "Pepper" Martin on his wing. Lt. Enrique Provencio was flying Daddy White Three with Lt. Ken Hart bringing up the rear. Champlin and Provencio had already shot down one Zero each, and Hart was ready to claim his fifth when Daddy, White Three and Four became separated from the flight and moved in on Daddy Green's prey.

Hart got into position when the Oscar was flying to the southwest at 6,000 feet. With a short deflection shot from the rear, Hart saw a flash of flame on the right side of the engine and the Oscar rolled over to crash east of Olango Island. With three other victories scored on November 24, plus one scored on December 2, Hart was now an ace. He added another Oscar during the December 7 patrol to make it an even six.

The 475th had claimed twenty-six victories on December 7, 1944 to surpass the 450th confirmed claim since it had started operations in August 1943. It also suffered the only command loss of a pilot when Lt. Colonel Meryl M. Smith did not return from his last combat mission.

Colonel MacDonald had taken off with Colonel Smith as his wingman for their first mission at 11:15 in the morning. They meant to join up with Clover White Flight, but contacted three Jacks first. The two P-38 aces immediately took the advantage and shot down two of the Japanese very quickly.

MacDonald got into a wild dogfight with the third enemy plane and Smith got in a good shot to relieve his leader by sending the Jack down in flames. The P-38s patrolled until a few minutes after twelve and returned to Dulag.

About two hours later MacDonald led another patrol with Lt. Leo Blakely as his wingman and Smith leading the second element. They were over western Leyte when a number of fighter identified as Jacks jumped them from the rear. MacDonald again turned the fight to the American advantage

John Tilley in the cockpit at Leyte. He flew McGuire's wing during the last of 1944 and regretted the experience. He got his fifth kill near Clark Field on December 26, 1944 in spite of difficulties offered by taskmaster McGuire. (Tilley)

McGuire and his Orderly Room staff. (Krane Files)

MacDonald again turned the fight to the American advantage by turning into the attack. One of the Jacks fell to MacDonald and another to Blakely.

Another Jack was on a firing pass at one of the P-38s, but Macdonald was able to get a good angle and shot most of the enemy's tail off, sending it tumbling into the water. Other Japanese fighters joined in the fight, but didn't manage much other than to get one of their number shot down by Blakely.

431st Orderly Room rigged for maximum ventilation. (Krane files)

Sometime during the fight Smith was attacked by two Jacks (or Georges of the Air Group 341). His P-38 was last seen heading for Ponson Island, away from the protection of his flight. Some damage from enemy fire was seen on his plane before he disappeared with Japanese fighters after him. With nine victories to his credit and a tour as group commander as well as yeoman service in various other positions in the 475th, Smith was a serious loss, indeed.

The ace race between Bong and McGuire went into a brief decline after the middle of December. McGuire and Lt. "Pete" Madison were on a sweep over Negros Island when McGuire got a full-deflection shot that shot down a Jack (or Frank, which type was also based in the area) for his thirty-first kill. After that date, McGuire is listed as being out of action with one of the serious diseases that he and most veterans of the southwest Pacific contracted.

Bong quickly raised his score to forty confirmed aerial victories by December 17. He was finally ordered out of combat and was on his way home by the end of the month. McGuire's boiling point would reach its zenith at the same time.

Dangerous Operations

Aside from those regrettable losses of pilots in combat there was also the inherent danger of operational accidents. The simple act of taking off in the accelerated tempo of December 1944 Leyte air missions had its own terrors.

Some groundcrews of the 431st Squadron escaped a near thing when an aborted takeoff by a P-47 on December 4 wiped out four of their P-38L-1s. The big Thunderbolt mushed into one of the P-38s parked along Dulag strip and the whole twisted mass of metal continued on and mangled the other three Lightnings. All four P-38s – and presumably the P-47, too, were transferred to the 10th Service Squadron.

On December 12, Lt. Bob Schuh of the 432nd Squadron was trying to land when the nosewheel of his P-38 became stuck in the up position. He was obliged to crashland with wheels up and his P-38 was transferred as a total loss to the 479th Service Squadron. Lt. Austin Morris of the 43rd squadron had the opposite problem when he was returning from a B-24 escort on December 23 and his nosewheel collapsed on landing at Dulag. Both Clover pilots escaped injury.

On December 11, Lt. Jack Purdy was leading a flight of four Possum P-38s on a PBY escort to Ormoc Bay when he ran into more than twenty Japanese fighters over one of their convoys. Purdy immediately went to the attack and shot down two Oscars while his wingman, Lt. Carl Redding, got another.

McGuire sometime near the end of his career when he became group Ops Officer and maintained a shadowy influence over the 431st Squadron. (Jeschke)

Lts. Miller and Hulett were in the second element and managed to claim one Oscar each.

Purdy finally decided to break off combat when fuel began to run low. However, the weather proved to be uncooperative when it closed in and forced the Possum flight around the southern tip of Leyte. Purdy had used up lots of gas in getting the two Oscars (his fourth and fifth victories) and ran out of fuel. He turned over the flight to Miller and crashlanded on Cabugan Grande Island.

Fortunately, he was picked up by Catalina in the afternoon and brought back to Leyte. The next day he got a ride back to the 433rd in an L-5.

Christmas Over Clark Field

By the middle of December the American forces were fighting from cramped quarters on Leyte and newly won ground on Mindoro Island. The Japanese were still in good position on the Luzon airfields, which locations were temporarily giving them the advantage of attacking American sea and air targets with impunity.

On December 13, Colonel MacDonald was obliged to dive through friendly antiaircraft fire from American ships below to get a Sally bomber making a run on the shipping. He braved return fire from the bomber to first shoot out the bomber's defenses and then set the aircraft aflame to crash

near a cruiser. MacDonald landed back at base after dark thoroughly exhausted from the necessary effort.

The first raids on Clark Field of any significance began on December 22. Even so, the Fifth and Thirteenth Air Force B-24s were hampered by the crowded conditions on Tacloban's airdrome, and were forced to bomb on alternate days. P-47s of the 348th Fighter Group escorted the heavies on December 24 and came home with a generous bag of thirty-three claims.

It was obvious that the moment was at hand to reap a harvest of aerial kills over the Clark Field complex. Whatever problems were being experienced by the bombers, the Japanese were obviously committed to put up a heavy resistance.

Tom McGuire was especially eager to get into the action over Clark. The 475th was scheduled to escort the bombers over Mabalacat, a field on the northern part of the Clark group, on Christmas morning. Sometime in December McGuire turned over command of the 431st Squadron to Captain Bob "Pappy" Cline and assumed duties as Group Operations Officer. Documents carry the signature of Major McGuire as commander of the squadron as late as the end of the year, but other notes in the official history of the group suggest that Cline was commander as early as December 2.

On December 25, however, McGuire was at the head of the squadron when it set out by ten minutes after eight in the morning. Fifteen 431st P-38s rendezvoused with the B-24s an hour later over the southwest tip of Masbate Island.

Once again Colonel MacDonald was leading the mission at the head of Clover Squadron. He was leading a flight of three P-38s with Captain Danforth Miller as his wingman. He followed the 431st into the Mabalacat area by a few minutes.

McGuire entered the combat area at 10:35 with his formation flying at 21,000 feet to provide top cover for the bombers. The Japanese first came along from 26,000 feet and to the rear of the P-38s.

Gronemeyer got his last kill over Dinagat Island, between Leyte and Mindanao, on December 6, 1944.

Under less stressful conditions (e.g. the tightening ace race) Bong and McGuire would have been more relaxed with each other, but McGuire was driven to become the top ace and Bong must have been at the end of even his cool nerve. (Krane files)

The enemy planes – mostly Zeros to McGuire's reckoning – were actually committed to coordinated attacks before McGuire reacted. One Zero attacked the number four man of McGuire's flight and the intrepid P-38 ace made a great lufbery loop to get a deflection shot that scored hits around the wingroots. The number four man, Lt. Jim Moering, was glad to see the Zero burst into flames and blow up.

McGuire made a head on pass at another Zero that disappeared after the two planes had crossed paths. Before he could return to the bombers McGuire noticed two more Zeros below and went after them. Lt. Alvin Neal was McGuire's wingman and, remarkably, was able to stay with him during the attack. He saw his leader's target begin to shed pieces and emit smoke before it crashed near the town of Mexico.

McGuire's flight was down to 15,000 feet now, behind the bombers which were receiving aggressive attention from Japanese interceptors. Four P-38s were jumping two Zeros and McGuire saw three more Zeros come down out of the overcast onto the tails of the American fighters. In spite of yet two more Japanese coming down on his own flight, McGuire went after the Zeros closing on the other P-38 flight.

Firing from dead astern to about 60 degree deflection, McGuire hit one of the Zeros that marked a trail of smoke as it went down to crash near the town of La Paz. Major George Dewey of group Headquarters was along on the mission and witnessed the Zero crash.

The two remaining Zeros displayed their fabled maneuverability by looping around to the rear of McGuire's flight. McGuire then displayed his fabled skill by breaking away from and losing them. Another pair of Zeros tried to avoid a subsequent attack by McGuire's flight, but he was able to easily get on their tails. Unfortunately, only one of his guns would fire and he directed first Lt. Tom Oxford and then Lt. Neal to shoot them down. Both Japanese planes were seen to go down smoking and crash.

Jim Moering also shot down a Zero. Other flights of the 431st raised the score with Lt. "Rabbit" Pietz and Louis DuMontier getting two each, Pietz, two Jacks and DuMontier, two Jacks. Fred Champlin scored his eighth victory and John Tilley his fourth. Hal Gray got a Jack that was seen to crash and burn by Moering, and John Ballard and "Pepper" Martin each got a Zero seen to crash and burn.

OPPOSITE: Famous publicity photo of Bong and McGuire during December 1944. Bong was at the end of his tether, also, and couldn't tolerate McGuire's nervous taunts. (USAF)

The 431st also suffered the only casualties of the mission when Lts. Provencio and Koeck were last heard over the radio that they were going down near Clark. Provencio was heard to tell Koeck to bail out before he also radioed that he was trying to get as far from Clark as he could to abandon his P-38. Floyd Fulkerson was also missing at the end of the day, but was the only one of the three to eventually make it back.

A few minutes after the 431st engaged the enemy, Colonel MacDonald arrived with Clover Squadron at 19,000 feet. North of Mabalacat Field about twenty Jacks and Zeros suddenly appeared below at 14,000 feet and MacDonald immediately took his flight down to the attack. By using the P-38's superior speed and by constantly turning into the Japanese attacks, the Americans were able to split the Japanese formations and do the most damage.

The fight had worked its way down to about 7,000 feet when MacDonald got a good stern shot at a Jack. Closing to about fifty yards, the P-38 spat out shot and shell that tore large chunks of metal into the air from the Jack's fuselage. The Jack burst into flame and carried its pilot down to destruction.

McGuire covering the Wheel Inn sign sometime at the end of 1944. He was pressing hard and scored seven victories in two missions, getting a total of ten for the month of December 1944.

Up to that moment MacDonald had taken many high deflection shots with little effect except for one Jack that seemed to be falling out of control when it was last seen. MacDonald's number three man went out of the fight with engine trouble and Joe Forster came along to take his place just as MacDonald was taking after another Jack over Clark.

Closing on the Japanese fighter, MacDonald noticed what he took to be a phosphorus bomb under one wing. He fired a three second burst that started the Jack blazing from nose to tail. Just as MacDonald passed over the wreckage of his latest victory, the phosphorus bomb exploded and scattered wreckage over the east side of Clark.

Joe Forster had started the mission as Clover White Three with Dick Kimball on his wing. When Colonel MacDonald started the attack, Captain Henry Condon had led White Flight down in a spiral and shot down a Zero from which the pilot bailed out. Forster got in a full deflection shot at another Zero which Kimball saw burn and crash.

Several other Japanese fighters managed to get above Clover White Flight and split it up. Forster was able to drive two Jacks attacking Clover White One, but then had to run for it when another jet-black Jack came after him. Finding himself alone, Forster quickly joined up with Col. MacDonald in time to see him shoot down his second Jack.

Just after Forster joined up with Col. MacDonald a pair of Zero model 52s were sighted diving through the overcast and heading west. MacDonald took his flight in a fast climb in the

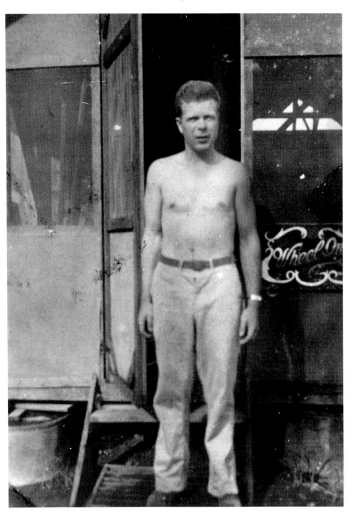

Dick Bong by the Wheel Inn barracks during his stay with the 431st when he scored his final four victories.

"Pappy" Cline probably sometime late in 1944. He ran the 431st when McGuire became the group Ops Officer in December 1944. Cline was rated as one of the best pilots and officers in the squadron. (Krane files)

opposite direction to get above these two enemy fighters. Apparently they didn't realize the P-38s were there and getting the advantage on them. MacDonald did a steep chandelle, ending up directly behind and a little below. At about 5,000 feet over the southern outskirts of Clark the P-38 put a long burst into the Japanese wingman and MacDonald watched pieces fly off and flames begin to trail along the Zero's fuselage. The pilot bailed out and MacDonald turned his attention on the leader.

Completely alerted now, the lead Zero pulled around into a tight turn. MacDonald used his dive flaps and got the proper lead, but excessive gravity forces jammed his guns and the Zero escaped. The antiaircraft fire was coming up from everywhere below and fuel was getting low, so MacDonald led his flight out of the combat and headed for Mindoro to refuel.

Lieutenant Harrold Owen claimed two Zeros in an hour long battle, and Lieutenant Lowrey Noblitt rounded out Clover Squadron's victories at nine Japanese aircraft. Possum Squadron was on a low level sweep over Mabalacat at the same time when a single Zero attacked Possum Red Flight. Lt. Glenn Maxwell was flying Possum Red Three and got within twenty-five yards of the Zero's rear to shoot it down in flames southeast of Mabalacat.

When the 431st retired from the battle scene at about noon they had claimed twenty-eight victories for one of their best days of the war. Three P-38s had been lost, but one of those pilots was eventually recovered. Somewhere in the process the group's 500th confirmed victory was recorded and Captain Henry Condon was the 432nd's newest ace. They were now Satan's Angels, indeed.

The past six weeks had been hectic, to say the least, so it was not surprising that little celebrating was seen in the 475th camp that night. Pilots and crews were well aware that much more combat was coming their way before Japan capitulated. As for Tom McGuire, he was content with the promise of action. His score now stood at thirty-five confirmed and the 475th history comments that he turned in early on Christmas night, "eager to make the most of the good hunting while it lasted."

John Pietz's P-38 at what must have been, according to the victory markings, very late in 1944. P-38J-15, serial 42-104034 could certainly have operating in December of 1944, but it is more likely that the victory marks were for the P-38 and that the photo was taken in mid-1944. Note the open recognition light panel behind the obviously proud crewman.

On December 26, another escort to Clark was mounted by the 475th. The past few days had taken their toll of the Japanese air defenses in the area, greatly curtailing both the number and aggressiveness of the interceptors. Some of the American combat reports for the Christmas day mission mentioned that the Japanese seemed almost certainly intent on ramming. On the next day they would just as determined, but not nearly so numerous.

December 26, 1944 did not begin as an unusual day on the air base at Dulag. The pilots and crews were up at the usual time and greeted a day that was dawning with the customary billowing white cumulus clouds on the horizon. After briefing and breakfast the pilots were ferried out to their waiting P-38s.

Tom McGuire would once again lead the 431st Squadron as Daddy Red One. His wingman would be Captain Ed Weaver, Fred Champlin would lead Daddy White Flight and John Tilley would bring up the rear with a two-man Daddy Blue Flight, Lt. Bo Reeves flying as his tail-end Charlie wingman.

Ten 431st P-38s were off the ground by ten minutes after eight and rendezvoused with the B-24s over Masbate Island at

nine o'clock. There was no incident throughout the run to the target while the 431st P-38s shepherded the bombers at an altitude of 16,000 feet, somewhat above and to the right of them.

While the bombers were heading away from the target McGuire saw five Zeros coming down on the rear of the formation from out of the overcast. The P-38 ace lost no time in diving after the enemy interceptors and got behind one that was pressing home his attack rather than breaking away. A single 45 degree deflection shot from about 400 yards away made a direct hit on the cockpit. The Zero continued to drive on the tail of a B-24 and McGuire closed to within thirty yards to fire another burst. Weaver saw this Zero explode.

McGuire happened to be in the heart of the enemy interception and took advantage of the opportunity. Weaver managed to stay with the wild-flying McGuire and saw him take another shot at a Zero with about 60 degree deflection which flamed the Japanese fighter. Three more Zeros came in to join the fight and McGuire got another good 45 degree deflection shot that sent one of them crashing into a dry stream bed.

Lt. Christopher Herman was leading McGuire's second element and having a hard time following McGuire while shaking Zeros off his own tail, but managed to see this third victory catch fire and crash. McGuire had by this time lost Weaver, although Herman was able at least to keep his flight leader in sight. He was able to confirm McGuire's fourth claim of the day when the ace chased yet one more Zero into a dive and made another remarkable deflection shot to send his target burning and crashing near the same dry stream.

With his steam now definitely up, McGuire found a Tojo in the distance and was excited at the prospect of getting five kills in one combat. He went after the Japanese fighter, which was now climbing for the protection of the overcast. Other P-38s were slipping down from above to give McGuire some competition for his prey.

Lt. Sammy Pierce was a member of the 8th Fighter Squadron flight that happened to be in the same area when McGuire started after the Tojo. He watched in mocking disapproval while McGuire and every other P-38 pilot who tried to get the enemy plane muff one firing pass after another. The Tojo would be in the clouds before any of the overeager Americans who seemed to be climbing over one another could make a decent shot.

When the Tojo seemed to be just about able to make the safety of the sheltering mist, Pierce shook his head and fired one good burst in a dive from his position above, and sent the fighter down in flames. Major McGuire filled the air with the rudest suggestions about the virtueless thief from the 8th

Squadron, but Pierce went home satisfied; he now also had four victories for the mission.

Three other 431st Squadron pilots scored during the mission. John Tilley chased one Zero through clouds over the tops of some mountains and finally shot it to pieces over the west coast of Luzon and watched it crash for his fifth confirmed kill. He also saw Fred Champlin shoot down a Zero for his ninth kill. "Bo" Reeves shot down a Zero three miles south of Clark, witnessed by Champlin at the same time.

Captain Jack Fisk was leading two flights of 433rd Squadron P-38s over Clark when he saw McGuire's flight engaging the Zeros about a thousand feet above. Possum Red One (Fisk) led his own and Possum White Flight up into the battle.

Unfortunately, Fisk had trouble with his engines and by the time he got them running properly White Flight called out an attack from the rear. Fisk turned into this aggressive Japanese fighter, which was identified as a Tojo, and convinced its pilot to break off and dive away. The American followed and got in some good hits around the cockpit. The Tojo went straight down and crashed just southeast of Clark strip number 5.

That mission ended the Satan's Angels scoring for the year. They had confirmed no fewer than ninety nine kills in December and the total now stood at 522. Opposition over Clark was disintegrating with the effectiveness of American fighter missions.

With the end of the year came a welter of rotations home. One of the old hands of the 432nd to go home in December was TSgt Carroll Fithian, who had been Jim Ince's crewchief at the beginning of 475th operations. 433rd pilots to go home by the end of the year included Bob Tomberg, Herb Cochran, Bob "Andy" Anderson, Joe Price and Don Weimer.

By the end of 1944, Dick Bong was also on his way home as the leading American fighter ace of all time. He had scored five of his victories with the Satan's Angels and had left a permanent mark on the history of the group.

Tom McGuire both resented and admired Bong. He was determined to wrest the title from his rival, but also realized that the time was short. Fatigue and the effects of tropical illnesses would soon end the combat career of this fearless ace. Whatever he did would have to be done soon and McGuire was pushing the issue to the limit.

Cline's #135. Bong flew this P-38 when he operated with the 431st in December 1944. (Krane files)

DOOTS II with pilot Tom Oxford on right. Original DOOTS was probably retired after McGuire's fatal mission of January 7, 1945.

Chapter XIV

LUZON AND THE INDO-CHINA COAST JANUARY-MARCH 1945

N
ineteen forty-five started out in a lively way for the Satan's Angels. After a brief lull at the end of December the 475th returned to Clark and claimed another seven Japanese aircraft shot down, with an additional two probably destroyed and one damaged.

All three squadrons were off by 8:30 in the morning with a relatively small total force of fifteen P-38s. They rendezvoused with the bombers at fifteen minutes to ten. Col. MacDonald was again leading the group at the head of Clover Squadron's five P-38s. Campbell Wilson was leading the three planes of Possum Red Flight.

Clover was the first to get into action when it reached the target area at 14,000 feet, just under the overcast. MacDonald sighted a lone Dinah coming in from the northwest at 12,000 feet and immediately went after it in a head on dive. The reconnaissance plane turned to get away from the P-38 and MacDonald was able to fire only a short burst at about 40 degree deflection. It was enough, for,when MacDonald turned to make a stern attack, he saw the Dinah in flames and rolling over into a vertical dive to disintegrate below.

When MacDonald looked around for other Japanese aircraft, all he saw were P-38s shooting them down. It would have been another field day except that there were very few enemy interceptors in the air and they were not as eager as the pilots of just a few days ago.

One Tojo camouflaged in very dark green, shiny paint (perhaps from the 246th Sentai, which would likely be operating in the Clark area) was spotted at 4,000 feet flying toward the west. MacDonald led Clover White One, Captain Paul Lucas, down to the attack on the stern of the fighter.

Apparently, the Tojo pilot was completely unaware of the presence of enemy P-38s on his tail. A series of short bursts set him afire and Lucas witnessed the plane crash about ten miles east of Clark Field. It was Colonel MacDonald's twenty-sixth victory, putting him in the ultra-select band of American fighter aces to reach that mythical total.

At about this point, Lucas lost contact with MacDonald when the colonel's P-38 passed under a cloud layer. Three Japanese fighters that were later identified as "Zeke 52s" passed overhead from a ninety degree angle to Clover White Flight's path. Lucas split his flight and joined newly-promoted Captain Chris Herman and Lt. Douglas Thropp of the 431st Squadron to give chase.

Lucas got on the tail of one Japanese fighter and closed to within fifty yards, getting strikes all over the left wing and fuselage. The right landing gear dropped and Doug Thropp watched the flaming wreckage crash while Lucas unsuccessfully chased the now alerted lead Zero.

Chris Herman chased the third Japanese fighter above the cloud layer and shot it down. He was the only 431st pilot that day who managed to get close enough to score an aerial victory.

Major Edwin Weaver by his #102. He flew #138 in the 431st Squadron on occasion. #138, P-38L-5, Serial 44-25798 was flown by Lt. Bill Bolinger, and F/O Tony Paplia at the end of the war. (Weaver via Krane)

When the went below the cloud layer Colonel MacDonald had seen a parachute descending and decided to investigate. His combat report contains an unusually incidental comment: "It turned out to be an extremely nonchalant Nip, in a dark shiny green flying suit, floating down with his legs crossed and his elbow propped up on the shrouds. I returned to home base via Mindoro . . ."

The major fight of the day was experienced by Possum Squadron. Captain Wilson was leading Possum Red Flight at 19,000 feet when he noticed a large formation of fifteen Jacks below. The P-38s let down in a great spiral onto the tails of five Jacks with generally wild results.

Wilson damaged one of the Jacks and his wingman, Lt. Edward Edesberg, damaged a second that was finally shot down by Possum Red Three, Lt. Bert Roberts. Possum Red and Possum White Flights then became separated and Wilson flew a complete circle while he pulled up from his attack after which he attacked another formation of Jacks.

One of the Japanese wingmen was a little slow in evading Wilson's bursts from 500 to 100 yards and was obliged to parachute from his burning fighter. While he was trying to find the rest of his two flights Wilson spotted a Nick coming through the clouds slightly above. The P-38 pilot deftly chandelled onto the tail of the blue/gray twin-engine fighter with its distinctive green mottled camouflage. With two long bursts from the P-38's guns, the Nick began to trail smoke and fire from its right engine. Taking a moment to observe his target, Wilson fired again from below and the engine exploded, setting the whole wing and right side of the fuselage aflame.

Wilson watched the ball of fire spin to earth just north of the great Laguna de Bay east of Manila. His combat report also

contained an observation about the enemy he encountered that day: "Enemy fighters observed were flying excellent formation, but seemed somewhat at a loss as to just why. All enemy fighters seen by me were being harassed and destroyed by our P-38s and remained on the defensive exclusively."

The observation was valid. With one defeat after another being inflicted on them, the Japanese were forced to make the most conservative defense preparations. It was no longer a matter of if but when the Americans would reconquer the Philippines.

Ironically, the American forces had almost duplicated the Japanese successes at the beginning of the war. Japanese forces first neutralized American seapower, then took command of the air before making landings in Lingayen Gulf on December 22, 1941. The American return happened almost exactly three years later on January 6, 1945 at the same strategic location less than 150 miles from Manila.

John Tilley got his last victories at the end of 1944, usually flying wing for McGuire. Although he greatly respected McGuire as a pilot he did not enjoy the experience flying with such an overly demanding leader. (Tilley)

Left: McGuire before the VIP quarters of the 431st Squadron (WHEEL INN) late in 1944 or early January 1945. (Krane collection) Right: McGuire photographed on January 5, 1945; just two days before his death. (National Archives)

The Japanese had already decided to withdraw into the interior of Luzon, just as MacArthur had done in 1941. The Cagayan Valley to the northeast of Manila was ideal for Japanese tactics. American supplies would be restricted to the narrow Balete pass when it could be cleared. The limited Japanese forces could live off the land and retreat grudgingly.

Japanese airpower was already ground into fragments by the time Luzon was invaded. There were still operating units on Negros and a few other islands, but the prospect of major aerial engagements seemed to vanish with the landings on Luzon.

The Death of Tom McGuire

This dwindling opportunity for increasing his score tormented Major Thomas B. McGuire, second-ranking American ace of the war. He was tormented by thought that he would be pulled from combat just as he was about to realize his fervent goal to become the ranking American ace of all time.

431st Squadron mission number 1-668 a four plane fighter sweep to Fabrica Airdrome on Negros Island took off on January 7, 1945 at 6:20 in the morning. Major Tom McGuire was flying the P-38 that happened to be shared at the moment by Fred Champlin, who was scheduled to rotate home in the next few weeks, and Lt. Hal Gray; it was squadron number 112.

Captain Ed Weaver went along as Daddy Special Two, flying Tom Oxford's number 122, "Doots 11." Jack Rittmayer led the second element in Lt. Rohrer's number 128, and Doug Thropp eagerly accepted the invitation to come along as tailend Charlie, flying number 130, "Miss Gee Gee."

The four silver and red P-38s leveled off at 10,000 feet, on course for Fabrica. The undercast was solid below at 6,000

feet, so McGuire eased his flight down when they reached Negros and broke out below the clouds at 1,700 feet. The Americans were about ten miles north of Fabrica strip and boldly began to circle the base at exactly seven o'clock.

Five minutes later, McGuire set course at about 1,400 feet for the airstrips on western Negros, despairing of finding Japanese over Fabrica. Within a short time Weaver sighted what he took to be a "Zeke 52" climbing directly below about 500 feet and ahead about a thousand yards.

What Weaver identified as a Mitsubishi A6m-5 Zero was actually an Oscar flown by W/O Akira Sugimoto of the 54th Sentai. He had been flying in search of an American supply convoy headed for Mindoro or Lingayen Gulf. The weather had been impossible and Sugimoto headed back after a long and frustrating flight.

Weaver had called the enemy fighter and McGuire made a diving turn to the left to trap it; Sugimoto was already directly beneath the P-38 flight. The Oscar pilot may have been tired after the long search mission, but he was sharp enough to turn left himself, and get on the tail of Lt. Thropp, who was now the number three P-38 after being ordered to switch positions with Major Rittmayer.

Thropp skidded his Lightning to avoid the fire coming from Sugimoto's two 12.7 mm guns. It was inconceivable to Thropp that the Japanese pilot could miss but he did! Rittmayer put his P-38 on the verge of a stall to get enough lead to discourage this pugnacious Japanese. That he managed to do, temporarily. Sugimoto simply tightened the turn on his extremely maneuverable fighter and got a bead on Weaver.

With all he could to avoid the attack, Weaver called McGuire and tightened his own turn until he was inside and a little below his leader. Sugimoto stuck like glue in his light olive-green Oscar with its graceful yellow-orange tail insignia

Fred Champlin by his last P-38, EILEEN-ANNE, He shared P-38L-1, serial 44-24845, with Hal Gray at the time that McGuire flew it on his last mission. (Champlin via Krane collection)

Captain Paul Lucas earlier in the war. He had flown with the 432nd from the first days until he went down to his death on January 15, 1945. (475th history)

and confounded the entire flight of battle-tested Americans. A call from McGuire ordered the flight to keep its drop tanks even as the Oscar was out turning the P-38s. The tanks were nearly full and further hampered the flexibility of the American fighters. Within a moment that inflexibility would produce catastrophe.

Weaver saw McGuire "increase his turn tremendously" to get his sights on the Oscar. The last thing Weaver observed of his leader was when the P-38 "snap-rolled to the left and slipped in an inverted position with the nose down about 30 degree. Because of the attitude of my plane, I then lost sight

Captain Louis DuMontier (pronounced Doo - Mont' - eer) probably soon after he had claimed two air victories over Clark on Christmas day 1944. His unusually named MADU P-38s were a result of parting company with his latest girlfriend. He wanted to name his aircraft after the main lady in his life at the moment – who happened to be his mother – but didn't want to endure the jibes of his squadronmates, so he resorted to the relatively cryptic name he had uttered as a baby boy. (DuMontier via Krane)

of him momentarily. A second later I saw the explosion and fire of his crash."

Sugimoto either saw his opportunity to escape or was driven off by Doug Thropp who had come around in the circle sufficiently to fire a three-second burst at the Oscar. The Japanese fighter raced off to the north where it made a forced landing, probably from damage received by Rittmayer or Thropp, and Sugimoto was instantly shot to death by a group of Filipino partisans.

Meanwhile, Sergeant Mizonori Fukuda of the Ki-84 Frank equipped 71st Sentai was landing at Manapla strip on Negros when he noticed Sugimoto's plight to the north. He raced to the aid of his comrade in the Oscar and arrived just about the time McGuire crashed and Sugimoto escaped into the clouds.

While the three remaining P-38s were still in disarray, Fukuda dived from the clouds to the left and got onto Rittmayer's tail in the middle position. All the remaining P-38s had dropped their tanks and Weaver, in the third position, got a burst at the Frank just as it fired a killing shot at Rittmayer from 90 degrees deflection. A moment later another explosion was seen on the ground less than two miles from Pinanamaan Town. Fukuda had also put a cannon shell into Thropp's right tail boom and left engine manifold. Weaver had done some damage to the Frank which made it back to Manalpa and crashlanded with twenty-three bullet holes from Weaver's guns, and was a complete write-off.

For many years afterward it was believed that McGuire was lost in action against a single Zero flown by an extraordinarily good pilot (one who was able to fly in two directions at the same time!). The true story is that two Japanese Army Air Force fighters had claimed two veteran and skilled P-38 pilots, but at a cost of two of their own fighters and one spirited pilot.

431st flightline early in 1945. #121, Lt. O'Neill (left side: BONNIE) #125, Capt. DuMontier (left side: MADU IV) and #111 Capt. Hart (left side: PEE WEE V). Note that #125 has the demon head on the radiator while #125 does not. (Krane collection)

Whatever the result, Major Thomas McGuire the very embodiment of the Satan's Angels valor, was dead and 475th Group morale was shaken to a degree for the rest of the war.

Bombing and Strafing

Since the end of December the 475th had been experiencing a shortage of P-38s. Even the high-ranking Tom McGuire had to wait for a new P-38 when his PUDGY IV had been used up, hence the borrowed airplane on his last mission. The 431st Squadron was obliged to send up formations of a single flight of four during most of January 1945.

While the reduced numbers restricted the chances for aerial combat, they were suitable for missions like convoy patrol and ground attack, which operations were becoming increasingly necessary in the push northward on Luzon. More

often the Satan's Angels would earn the enmity of Japanese ground troops who learned to hate and fear the freely marauding P-38s that hastened their retreat toward Manila or the Cagayan Valley.

Part of the 475th shortage of P-38s was due to some of its air craft being transferred to the 8th and 49th Fighter Groups as a payment on account for all the advantages enjoyed in the past. The righteous indignation expressed so emphatically for so long because of the priorities expended in the 475th's favor had at last been heard. The Satan's Angels could not be dominated by the Japanese, but quietly submitted to the forces of command.

As it was, the commander of the 432nd Squadron was killed in action during one of the first strafing missions of the new year. Captain Henry Condon was a fair haired, blue-eyed gentleman of the south who was an extremely popular commander. When he took the lead of eleven Clover P-38s to

Chris Herman got his air victory for the 431st Squadron on New Year's day, 1945. He is displaying his radiator demon that was applied around the same time.

A group of 431st Squadron luminaries around the beginning of 1945. Back: John Tilley, Hal Gray, Fred Champlin, Ken Hart Front: Louis DuMontier, John Pietz, Horace Reeves, Merle Pearson.

Colonel MacDonald's PUTT PUTT MARU, P-38L-5, serial 44-25643, after it was damaged on January 27, 1945. It was apparently in service with the 475th just a very short time and then was reissued to the 80th Fighter Squadron after it was repaired. (Krane Collection)

Leo Blakely was part of the last combat generation of the 432nd Squadron. He got his aerial victory during the big December battles and flew with the group until the end of the war. (475th History)

cover B-25s on a strike to Porac and Floridablanca Airdromes early in the morning of January 2, 1945 he was at the height of his combat career.

The rendezvous with the bombers was made over Biliran Island, but contact was lost in the overcast and Condon led his P-38s to the target area. A few miles north of Manila the P-38s found a train filled with Japanese soldiers, and Condon led some of his Clover Squadron P-38s down on a strafing pass. The locomotive exploded on Condon's first pass, but Condon's P-38 began smoking as he was pulling around for another go at the train. The nose of his P-38 dropped and he was seen trying to get out of the cockpit.

Before he could escape, however, the P-38 dropped from 2,000 feet and exploded when it crashed into a field about a hundred yards north of the highway running through the village of Bignay. Captain Elliot immediately took command of the 432nd.

Clover Squadron flew its first and perhaps most disappointing bombing mission of the new year with a skipbombing strike at a bridge north of Manila on January 9. Four P-38s, each with a single 1,000 pound bomb, took off from Dulag at 8:15 in the morning. The bridge was already being heavily mauled by B-25s and A-20s, so the Clover flight went to an alternate bridge and dropped three bombs harmlessly, while the fourth was jettisoned by accident.

Weather hampered operations for the next few days, but the pilots took every opportunity to get at targets on the ground. P-38s were still in such short supply that a mission like that of January 13 had to cover a convoy off the island of Panay with two P-38s from the 431st and two more from the 432nd Squadron to make up a complete flight of four. The pilots noted that Silay strip on the northwest coast of Negros had about fifteen serviceable single-engine fighters on repaired runways.

Lt. Werner's #139 (Krane files)

Groundcrew by the tail of Champlin/Gray's #112. (Krane collection)

Captain Paul Lucas, another 432nd veteran from the early days of New Guinea operations, was leading a patrol of six P-38s on convoy patrol during the morning of January 15 near Negros. When the patrol was over he decided to take a crack at Silay. During a strafing pass, Lucas was hit by light antiaircraft and machine gun fire and was forced to crashland. His P-38 was unfortunately coming down much too fast and he crashed into a field on the west bank of the Malisbog River. Some Filipinos were seen waving white handkerchiefs and running to the aid of Lucas, but he was sitting slumped in the cockpit with the canopy open and the dust and smoke of the crash still swirling around the wreckage of his P-38. The five remaining P-38s landed at Dulag at 11:35 and offered little hope that he could have survived.

Another four P-38s from Clover Squadron made a return visit to Silay on January 17 and released four 1,000 pound bombs that failed to explode for some unknown reason. The frustrated pilots strafed the strip and revetment area, but couldn't be sure what damage was done.

Two days later the 432nd lost another pilot to much happier circumstances than the beginning of the month. Cap-

tain Richard Kiick had come to the Squadron fresh from the U.S. on September 30, 1944 and now was reassigned to take command of the 421st Nightfighter Squadron.

The most successful bombing mission flown by the 432nd Squadron was a strike at the airstrip near the resort town of Baguio, north of Linguyen. Four P-38s – three with a single 2,000 pound bomb and the fourth with a thousand-pounder – took off at fifteen minutes to 11:00 in the morning and arrived over the target at 1:30. The three heavier bombs landed in the middle of the strip and the fourth exploded just off the west end of the runway.

This January 22 mission was not only the most successful of the month for the 432nd, but the 431st also bombed the strip well with a flight of P-38s. All their bombs fell within the target area, making the Baguio strip unusable to counter the widening Lingayen landings.

The 433rd Squadron had been performing yeoman service during January on PBY escort, courier missions and convoy patrol. During one of those convoy patrols on January 24 the last Possum Squadron aerial engagement and victory was recorded.

Hal Gray by his #112, serial 44-25656. Apparently tail is removed, either for maintenance or damage repair. (Krane collection)

Bob Weary by Gray's KIM IV. (Krane collection)

Lieutenant LeRoy Ross was leading a flight of four P-38s over the convoy south of Sequijor Island to the east of Negros. About forty miles south of Sequijor Ross sighted two Zeros in very dark green camouflage coming directly from the side at the same altitude of 9,000 feet.

The Japanese fighters retained their droptanks and seemed either unaware or uncaring of the presence of the convoy and escort. Ross ordered his flight to jettison tanks and quickly got into a firing pass on the tail of the leading target. With a short burst of fire, Ross set the right wing and right belly tank* of the Japanese fighter on fire. The pilot bailed out and his aircraft crashed about eighteen miles off the coast of Mindanao. The other Japanese plane escaped.

On the morning of January 27 Major John Loisel took off in Colonel MacDonald's P-38 PUTT-PUTT MARU, leading a flight of four 432nd Squadron P-38s to patrol a convoy south of Leyte. When the mission came to an uneventful end the Clover flight came down to land at Dulag by 10:45. Loisel set

down on Dulag strip number one and stopped at the end of the rain-soaked and congested runway to talk down the rest of the flight.

His P-38 was turned in the direction of the incoming flight and he felt a vague uneasiness when one of his P-38s came straight at him a bit too fast and a bit too long. The runway was still not good for anything but an ordinary landing, and Lieutenant Arnold Larsen was coming in hot and flat. Before he knew it he was on the slick surface of the runway heading for PUTT-PUTT MARU.

There was nothing he could do about it before Arnold skidded head on into Loisel's stationary fighter. The last thing Arnold saw before the embarrassing crunch of the impact was the surprised and reproving face of his leader. Nobody was injured, but both P-38s were sent off to the 10th Service Squadron. Lt. John Berry of the 433rd Squadron also collided with a 432nd P-38 on the same rainy strip and both aircraft were taken out of service. PUTT-PUTT MARU was repaired and subsequently assigned to the 80th Fighter Squadron.

---

* NOTE: The fact that the "Zero 52" had two belly tanks and that one of the aircraft had a broad orange around the fuselage suggests that it may actually have been a Ki-84 Frank of the 22nd Air Group.

Apparently an earlier photo of Oxford's P-38 before radiator motif was applied. (Krane collection)

Philippine Notes

Throughout the end of January and the beginning of February the shortage of P-38s continued. Only two or four plane flights could be mounted due to the lack of aircraft in the 475th. Missions were more often transport escort or routine patrols, the more intensive operations being carried on by other units on Mindoro Island.

Some contingents of the 475th moved up to Mindoro around February 3 and set up camp with the 8th Fighter Group. The Satan's Angels pilots and crews were delighted to get back into some of their own P-38s while other crews and aircraft were flown up within a short time.

The Satan's Angels were based on Elmore Field which proved to be one of the worst ever used by the group. Located between San Jose and Bugsanga on the southwest tip of the island, Elmore was in the middle of a treeless, brown plain which was baked airless by the sun. Great clouds of brown dust blanketed everything and the green mountains to the north were beautiful, but blocked every whisper of breeze except those coming from the distant sea.

Cline's #135 after the radiator motif application. (Krane collection)

Cline by his P-38. (Krane collection)

At any rate, the 475th was stronger and spoiling for action now that it was nearer to the real fighting. Missions were flown to search for enemy convoys, sweep the Cagayan valley where the Japanese were slowly withdrawing in the face of overwhelming American pressure, and divebombing those enemy positions that withdrew too slowly.

Perry Dahl led one of the divebombing missions on February 8 to Meriveles on the Bataan Peninsula. MacArthur had preempted a Japanese repeat of his defense of the peninsula by making a landing there, thus sealing the Japanese into a secondary retreat toward Manila.

The four P-38s that Dahl led each carried a 500 pound bomb and a belly tank. They successfully bombed the initial target around ten o'clock in the morning and Dahl led his flight down to strafe Mariveles barracks and a jetty at Cochinos Point. It was not air-to-air combat, but the pilots were beginning to feel as if they were back in the fighting.

Jim Moering by #124 after name IRISH was deleted.

On the same day nine group pilots flew back to Hawaii to pick up brand-new P-38s. Other fighters were being sent up from Biak to bring the 475th nearly up to full operational strength.

Satan's Angels Over the China Coast

Enough P-38s were received to permit a sixteen plane escort of B-24s and B-25s over a Japanese force of ships along the South China Sea coast on February 13. Colonel MacDonald once again led the 475th escort with a five plane flight from the 432nd. They took off from Elmore Field at 7:30 in the morning.

The bombing was unsuccessful because the American bombers were obliged to salvo their loads through a heavy undercast. Heavy warships in the Japanese shipping precluded an effective low level run and the escort was headed home by 11:45.

"Pete" Madison by Cline's P-38 sometime early in 1945. (Krane's collection)

Ten minutes later a single Ki-57 "Topsy" a Mitsubishi civil transport that was pressed into service much as the Douglas DC-3 that became the C-47 was sighted at 10,000 feet headed for Hong Kong. MacDonald started his curve of pursuit and could plainly see the graceful curves of the distinctive enemy plane in its mottled green camouflage. Several bursts of gunfire from 40 degree deflection to dead astern set the Topsy afire and some of the passengers were seen to jump without parachutes to escape what had become a flaming coffin for them.

Another fighter sweep to Formosa signalled the 475th Group's continued presence along the China coast on February 16. With other missions flown on ground support in the Cagayan Valley as well as other tactical locations, The 475th was fully employed once again in the operational agenda of General Kenney's Far East Air Force.

Firebombing missions began at the end of February. Joe Forster led an unusual firebomb mission on February 26 to deny a crashed B-25 to the Japanese. By the time the P-38s were finished ". . . a souvenir hunter couldn't have salvaged enough aluminum to make a decent set of earrings for his lady fair."

Last Air Victories

During one of the long range "Snooper" missions over the coast of what is now Vietnam the 431st scored some of its last aerial victories of the war. Twelve 431st Squadron P-38s took off at 7:25 on the morning of February 25 and rendezvoused with a pathfinder B-25 five minutes later over the beach south of Elmore strip.

The P-38s never did make visual contact with the PB4Y "Privateer" they were supposed to cover, but found Japanese shipping in the Cam Ranh Bay area and went down to strafe. One metal lugger was destroyed and three other ships took some probable hits. The best discovery, however, was a seaplane base with several A6M2-N Rufe floatplane Zeros anchored on the water. Four of the Rufes were destroyed and another damaged.

Two of the Rufes happened to be airborne and decided to take on the whole flight of P-38s. Lieutenants James Barnes and Tom "Pepper" Martin shot down the outrageously intrepid Japanese, one of whom managed to bail out before his Rufe crashed on a spit of land jutting into the sea.

475th flight over the Philippines, late 1944 or early 1945.

Bob Weary flew 135 at some point, perhaps before Major Vogel took it over in June 1945. (Krane collection)

"Red" Herman and crewchief in the Philippines with #118, serial 44-26303, P-38L-5.

MISS BOBBY-SOX, pilot: Lt. Chester Parschall. (Krane collection)

Perry Dahl got the eighth of his victories on a fighter sweep to Formosa. Seven Clover planes reached the target (one returned early because of accidently released drop tanks) and began a strafing run on a group of sailing ships before Dahl ordered the attack stopped on what seemed to be civilian craft.

When he pulled up from his strafing run Dahl could see an unidentified plane at 1,500 feet heading inland. The bogie was quickly identified as a Sally bomber and Dahl went after it. He was closing quickly on the Japanese bomber when the pilot reacted to the presence of the American fighters and dived for the sea.

Just as the Sally began a turn to the left, Dahl fired a short burst that started its right engine burning. Two more bursts from the P-38 and the Sally went down with pieces coming

QUEEN "B" in the Philippines early 1945. In the background is #133, Bob Werth's IRISH ANGEL. (Krane collection)

No regular pilot is listed in 431st records as flying #129. However, this photo seems to imply the P-38 had a proud groundcrew and Lt. Merle Pearson returned early from a mission on January 1, 1945 in 44-24848, numbered 129.

Above and below: Bob Werth in and around IRISH ANGEL, P-38L-5, serial 44-25482, his P-38 from around the first months of 1945. (Krane coll.)

back from its fuselage and tail. Even though the bomber crashed on the beach and exploded, five survivors were seen running from the wreckage. This was Dahl's first fight since his ordeal of November 10.

If the 475th was sputtering in terms of aerial engagements at least it ended its days of victories with some flair. Dahl had scored his eighth victory in the first engagement from Clark Field bases. On March 2 the air echelons of the group began to move into the strips near old Fort Stotsenberg and the camps were quickly set up.

It was a much happier base that the 475th was building on the old Clark strips. There were real buildings, paved roads

Lt. Bill Bolinger who joined the 431st sometime early in 1945 and took over #138, MRS BILL, serial 44-25798, crewchief: Tony Paplia (Krane collection)

First Lieutenant James T. Miller, another pilot who came into the 431st around the beginning of 1945.

Bob Werth and groundcrewman around the demon head on the 431st commander's P-38. Parschall's #132 is in the background. (Krane collection)

Lieutenant Robert D. Harris by his P-38, probably after Howard Max had relinquished the number in 1945. (Krane collection)

Bob Weary, another latecomer to the 431st by the fearsome squadron logo. (Krane collection)

Ed Weaver's P-38 with score from North African campaign when he shot down two German aircraft and probably two others in P-40s of the 57th Fighter Group. (Krane collection)

Young Bob Werner in his P-38 after joining the 431st in the Philippines. (Krane collection)

and even a piped-in water supply. The group surgeon noted a VD trend not long after setting up camp on Clark, the first time such a thing happened in 475th history.

During March the last combat contingent of new pilots and crews arrived in the group. The next time a sizable number of crews would be assigned their duty would be primarily postwar occupation.

On March 6, in the early predawn hours, the Japanese greeted their old adversaries, the Satan's Angels, with an air raid. No antiaircraft defenses were in place, so the raiders

Jack "Foxie" Olson's #124, probably on Clark Field or Lingayen. (Krane coll.)

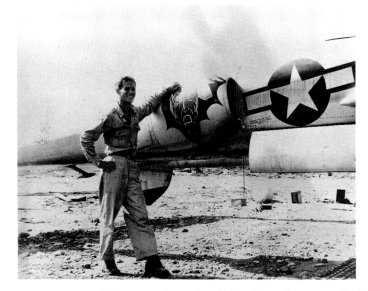

This page and opposite: 431st pilots and crews on Clark Field or Lingayen March-April 1945. (Krane collection)

caused more fear in stout 475th hearts than they had in months. That was the only damage, however, while two formations of bombers spread over ninety minutes failed to cause any material destruction.

Missions during the middle weeks of March consisted mostly of C-47 cover missions, patrols and an occasional escort to the China coast. On March 15, the 475th began ground support missions for American troops advancing along the coast north of Luzon and into the Cagayan Valley.

That day the 432nd Squadron sent twelve P-38s to the resort town of Baguio, just to the northeast of Lingayon Gulf, seventeen of twenty-two bombs fell near the primary target.

Another mission was flown to the same target by the same twelve P-38s. This time the bombs were dropped on Japanese personnel areas and a radio station. One building was left burning and the radio station was partially destroyed by a

direct hit. Some accurate light and medium antiaircraft fire was thrown up at the P-38s, but only the fighter flown by Lieutenant Lawrence Roberts took a hit from a single bullet in the wing.

During the latter part of the month, more missions were being flown to the Asian mainland, as well as Formosa and the southern limits of the Japanese home islands, themselves. After a few days of intensive divebombing missions northeast of Lingayen, in the general area of the Ipo Dam, the 475th flew one of its most important missions to what was then known as French Indo-China.

At eight o'clock on the morning of March 28, 1945, eight P-38s of the 431st Fighter Squadron, seven of the 432nd and four more of the 433rd took off on a B-25 cover mission. One B-25 would act as pathfinder to lead the twenty-one Satan's Angels to the Ben-Goi area. Major "Pappy" Cline had to leave

L-R: Cpl Edward M. Pierson, Sgt Patrick Antropik, TSgt Elmer Hines, SSgt Roy Paines, Cpl William Pappas. Cpl Robert Cronk sitting in background. 431st Squadron Armament Section February 1945. (AAF)

the 431st formation and turned the lead over to Lieutenant Ken Hart. Perry Dahl led the 432nd and Major Loisel led the entire mission at the head of the 433rd.

The P-38s found the Japanese ships, which included at least five destroyers and cruisers, but failed to make contact with the B-25s. Perry Dahl led his two flights down to a dangerously low altitude to orbit the Japanese convoy because of the haze that obscured vision. Black puffs of flak burst intermittently around the circling American fighters.

Dahl and his P-38s were at 7,000 feet when Banyan Red Flight* observed a few fighters below and dived to engage them. Twenty more Japanese aircraft were coming down from above and the radio was so jammed with excited calls that Dahl could not warn the P-38s already chasing the enemy below.

Japanese elements of two were breaking off the stacked formations and heading down for Banyan Red One; John Loisel on the tail of a Japanese Frank. Dahl pulled his P-38 up until it nearly stalled, but managed to divert the first of the attackers. The Hamps kept coming down at Dahl until he managed to catch one in a large angle deflection shot and it went down in flames.

Some of the Hamps were able to get through, however, and attacked Banyan Red Four. Dahl sent his second flight (Banyan Blue) down to help and they chased off the Japanese attackers, Lt. George Wacker getting credit for shooting down one of them.

While Dahl was busy deflecting various Japanese attacks coming from various sides, Loisel was shooting down a Frank. He attacked one which made a sharp turn in reaction to Loisel's sighting burst. The second Frank continued straight

---

* BANYAN was a special call sign for 475th formations of less than group strength.

ahead in a climb until it took hits in the fuselage and right wing, falling off steeply and trailing flames until it went into the water near the convoy.

Banyan Red Four, Lt. Wesley Hulett, had just been saved for the moment by Banyan Blue Flight and radioed Loisel to let him know that Hulett had one engine shot out. Loisel located Hulett in time to chase another enemy fighter off his tail. Hulett then radioed that both of his engines were out and Loisel told him to head due east where the Catalina rescue flying boat would be instructed to pick him up.

The 431st's lead flight got into the fight a little later than the other squadrons. Three Hamps jumped Ken Hart, "Pepper" Martin and "Bo" Reeves. Within a short time all three P-38 pilots had claimed victories; Hart forced one Japanese to bail out, Martin sent one down in flames and Reeves sent a lethal blast of shells into a third enemy cockpit.

Another flight of Hamps attacked the three P-38s after they had reformed. Hart broke sharply to the left and came out on the tail of one Hamp. Pieces of the Japanese plane came off from Hart's short bursts and the pilot was persuaded to bail out

after the Hamp started to burn. Looking over his right shoulder, Hart saw Reeves shoot down another Hamp that spun in just north of his own victory.

Lieutenant John "Rabbit" Pietz was leading the second element of Banyan Yellow Flight with Lieutenant James Barnes as his wingman when they dropped tanks after two bogies were called in at their level. The two Japanese fighters were chased to the deck where they were identified as Tojos when they turned to fight.

After some difficult turning, Pietz was able to close on the tail of one Tojo. Barnes was just completing a 360 degree turn when he saw Pietz firing close on the tail of the enemy plane before it rolled over and fell into the sea.

With nine more air kills to its credit, the 475th Fighter Group had broken the 540 mark. The only loss was, unfortunately, Wesley Hulett who was never found. The final air victories would come on the next day.

The 431st and 432nd Squadrons put up a total of ten P-38s for another B-25 antishipping strike on March 29. Taking off a little after seven o'clock in the morning, the P-38s rendez-

Olson and crewchief Richard Van Der Geest. (Krane Collection)

Above and below: Van Der Geest on either side of BOWLEGS II. Note the Satan's Angels T-shirt. (Krane collection)

Two photos at right: Foxie Olson by BOWLEGS II. (Krane collection)

431st Fighter Squadron sign on Lingayen and 432nd sign early in the Philippine campaign.

Pepper Martin scored his second victory on February 25, 1945 and got his third on the March 28, 1945 mission to the China coast.

John O'Rourke got the last 431st aerial victory when he downed a Zero over the Indo-China coast on March 29, 1945.

voused with the B-25 pathfinder over Capones Island and set off for a nearly four-hour flight to the target area.

One 431st and one 432nd P-38 returned early. Since the 432nd would be in the target area later than the 431st, Lieutenant John O'Rourke spontaneously volunteered to fill in the Banyan Green Four position.

Lieutenant Laurence LeBaron from Group Headquarters was leading the mission with Lt. Harrold Owen leading Banyan Green Flight. Lt. Lawrence Dowler headed the second element with O'Rourke happily bringing up the rear.

Around 11:30, as the flight was nearing the target area, LeBaron called out eleven Zeros at the same level as his own

Sign on Clark Field strip No.2, ca. March 1945.

Ken Hart got two Hamps over the Indo-China coast on March 28, 1945 for his seventh and eighth victories.

flight; 4,000 feet. Banyan Green made a climbing turn to the right and came down on the tails of the enemy. LeBaron immediately chose one of two stragglers and hit it in the fuselage and wingroot with cannon shots. The Zero began to burn and slipped into the water.

Lieutenant Owen fired at the other straggler which was seen to splash into the sea by Dowler. Owen then faced two more Zeros in a head on pass as they were closing on LeBaron. At this time, LeBaron heavily damaged another Zero and O'Rourke followed closely behind, so heavily engaged in shooting down his own target that he didn't notice LeBaron's Zero hit the water.

Dowler faced one Zero in a head on pass that resulted in the Zero passing overhead then going down to crash into the trees about a half-mile inland. Dowler then claimed a second Zero and LeBaron chased his third that was making for the clouds inland shot it to pieces.

This last combat ended prosaically with the P-38 flight patrolling for a short time then heading home around noon to land at 3:30 in the afternoon. Various victory lists give the 475th about 550 kills for the war. Whatever the final tally, the group averaged a remarkable twenty-two victories per month, the highest ratio in the Southwest Pacific.

For the next five months the 475th would soldier on in the role of ground support and patrol. The glory days were over, but the hard work of finishing the war was still ahead.

475th Fighter Group sign somewhere along the Pacific route. "In Proelio Gaudete": Be joyful in battle. (Krane collection)

Chapter XV

THE END IN THE PHILIPPINES APRIL-JUNE 1945

Strategic and Tactical

April 1945 was an especially cruel month for Imperial Japan and her war aims. The inner defense circle had already been cracked, leaving the home islands vulnerable to direct attack. On the first day of April, American forces landed at the very edge of Japan itself with the invasion of Okinawa.

Less than a week later, on April 7, the mighty and venerated Japanese Battleship "Yamato" sortied on a desperate and ultimately futile bid to counter the American thrust and was sunk. That same day B-29 bombers began escorted raids from the Marianas to Japanese cities, beginning a regular punishment of both the home island civilian population as well as its defenses. Things were little better in the Philippines. General Yamashita was being driven out of his luxurious Baguio headquarters with its swimming pools, golf courses and opulent buildings. His XIV Area Army was being driven slowly up the Balete Pass into the Cagayan Valley. By the end of April the Japanese would be out of Baguio and by the end of June would be divided into two major pockets, one to the northwest part of Luzon in the Sierra Madre mountains and the other fractured by American inroads, but concentrated in the Cordillera Central around the city of Bontoc.

To the south of Luzon, the Japanese were divided into the Shimbu Group being driven south of Manila and the Kembu Group which was isolated between Lingayen and the Bataan Peninsula. By the end of June these forces would also cease to exist as an effective fighting force.

April continued the accelerated tempo of the ground support role. Nine 432nd Squadron P-38s were loaded with Napalm or 1,000 pound bombs on April 2 and took off from Dagupan to bomb Japanese positions on the Villa Verde Trail, off Route #5 near Santa Fe to the northwest of Manila. The bombing was successful and the pilots agreed that the dreadful napalm bomb was perfect for the type missions they were now flying.

For the next few days the 475th would be diverted to long range strikes against shipping near Formosa, or targets on that island and the island of Hainan. On April 8, the Satan's Angels returned to Baguio and bombed concentrations of Japanese around the city. On the next day at least one of the missions pounded pillboxes southeast of the Balete Pass in spite of encroaching clouds that cut off visibility.

Throughout the first half of April it became routine to mount missions against the enemy positions around Baguio, the Balete Pass or even the Ipo area. Two or three missions a day were flown by the squadrons with an occasional escort or strike against the Asian mainland added to the agenda.

During the period April through June 1945 the 475th flew its heaviest mission schedule of the war. From the middle of March the number of ground attack operations increased until that type of mission became predominant.

431st sign at Lingayen in April 1945. (Krane collection)

Bombing and Strafing Again

One of the few things that 475th pilots enjoyed about the routine and alternately dangerous strafing and bombing missions was the chance to attack railroad locomotives. If the opportunity to meet Japanese in the air had diminished, at least there was some excitement in the prospect of racking up train engines which would sparkle with the impact of 20 mm cannon hits and spout tall geysers of steam to mark their destruction.

The 433rd flew an area cover mission for a B-24 strike on Shinchiku Airdrome on April 8 that gave the squadron an opportunity for railroad strafing. Six 433rd P-38s arrived over the airdrome at half-past noon and circled the area for fifteen minutes at 7,000 feet before breaking to the deck in search of targets of opportunity.

Just south of the city of Taichu, near the west central coast, a train was sighted on the main north-to-south rail line. The locomotive was turned into a hissing steam fountain by strafing passes and its five freight cars were thoroughly shot up. One of the cars was left in flames. The P-38s then finished the job by strafing a nearby railroad station and several freight cars on a siding.

A typical day for the 431st Squadron would be April 25. Ten P-38s were first off at 7:20 in the morning to bomb Japanese positions on a ridge northeast of Manila. White

Groundcrew of Reeves's EL TORNADO around April 1945. Reeves used #110 while commander Bob Cline used #135. (Krane collection)

SVENSKA was apparently a 432nd P-38, crewchief Sgt R.F. Paul. (Tabatt)

phosphorus smoke marked the target and the P-38s dropped two 1,000 pound bombs within the boundary. Twenty strafing passes were also made and the controller, who was apparently quite near to the target area, proclaimed the accuracy of the bombing with an exuberant call over the radio: "You're showering me with schrapnel and knocking me on my rear end, but I love it."

Later that afternoon nine more 431st P-38s took off to bomb a bridge near Bagabag, northeast of Baguio. One bomb was a direct hit that destroyed the bridge while seventeen others cratered the bridge approaches. The P-38s then strafed a Japanese camp area, setting fire to several buildings and hitting a gun pit on a hill southeast of the former bridge in thirty-six passes. All P-38s landed by 3:45 pm.

The 432nd had similar luck a few days earlier when it flew two missions to destroy bridges over the Marang River. The first was flown with napalm that failed to burn the structure. On the second try at another temporary bridge, Four P-38s skip-bombed the napalm tanks and knocked the span down with the raw weight of the projectiles.

That same day, April 18, the 475th got orders to move to Lingayen the next day. This bit of news sent morale sagging because the crews would have to leave the relative comforts of Clark for the crude airstrip and temporary structures of Lingayen beach. It was probably too good to be true, opined the older hands of the 475th, that the group would be long quartered on the more desirable grounds at Clark.

The reason for the move was that the Mustangs of the 35th Fighter Group could not operate from the Lingayen strip without excessive operational damage. P-38s, with their more rugged tricycle landing gear, were better suited to the rough surface of the Lingayen runway. Later, the strip would be composed of asphalt over steel mat, but was not so elaborate in April.

Its rugged landing gear was not the only advantage the P-38 had over the P-51 Mustang. The second engine was a valued safety factor in long over water missions. Lindbergh's contribution to P-38 flight endurance actually made the Lightning a longer-ranging fighter. Also, the P-38 proved able to get to altitude faster, making it a more desirable bomber destroyer.

None of these factors were applicable in the northwestern European fighting where the Mustang was supreme in the aerial combat role. However, the 475th Fighter Group, among other P-38 units in the Pacific, had proven the worth of the Lockheed fighter and the pilots and crews still revere the airplane.

All day on April 19 the trucks were busy shuttling between Clark and Lingayen, transferring the equipment of the

BOWLEGS II probably after an aborted mission in May 1945.

35th and 475th. There was little flying done between the start of the move and April 21, the disruption doing not much good for the mood and temperament of 475th crews.

The base at Lingayen was a mixed bag. While the tents of the 475th camp were near a beautiful white beach off the Gulf and flying time to the tactical objectives around Baguio was shortened, the sun was unmercifully brutal in the sparse shade and much work was needed to bring the squadrons up to operational form.

However, missions were being flown again by April 22 and Japanese positions around the slowly diminishing base of Baguio were being heavily pounded. The 475th was having good success with a combination of general purpose and napalm bomb loads and received a wealth of praise both from forward controllers and ground units involved in the advance around Baguio.

On April 26, two American divisions broke through into Baguio and the former Headquarters of General Yamashita was in the hands of his enemy. However, the Japanese were digging in all along the route into the Cagayan Valley and

433rd Squadron P-38 leading the rest off on a mission sometime during the closing stages of the Luzon campaign. (USAF)

American fighter-bombers would have plenty of work neutralizing those positions.

The next day Lieutenant LeBaron came down from group headquarters to fly the afternoon mission with the 432nd Squadron. Twelve P-38s took off at two o'clock to dive bomb enemy troops and supplies. Over the target the flights split up and bombed the areas around the town of Ipil.

LeBaron led the first of the three flights which set fire to a gasoline dump that seemed to turn one entire section of the town into a sea of flames. Two other gas dumps south of town were also set ablaze by the flight led by Lieutenant Henry Toll. Bob Schuh led the third flight west of town and demolished a series of buildings and underground shelters.

The entire area was then strafed with a concentration of attention on fortified pillboxes. When the 432nd P-38s withdrew, the pilots could see that virtually everything of use to the Japanese was either in flames or sending up billows of smoke and dust.

475th P-38s ended April with another series of raids against Formosa. The P-38s covered B-25s and B-24s throughout the day and went down on the deck to strafe truck convoys, airdromes, factories and barracks. By the end of the day on April 30, the Satan's Angels were more convinced of the wisdom of transferring the group to Lingayen.

Young Bob Werner near the end of the Luzon campaign. (Krane collection)

Hammond's Hike

On May 3, ten P-38s of the 433rd Squadron were on a divebombing mission to the Bontoc area of the Cordillera Central. Lieutenant Jerome Hammond was leading the mission and making a pass at two hundred feet in the vicinity of LaPante mine, Mankayan. when antiaircraft fire came up from the mine entrance.

An incendiary 20 mm shell struck Hammond's P-38, exploding in the left engine and causing it to catch fire. Hammond was flying much too low for a safe bailout and the escape windows were stuck, anyway. The only thing for him to do was turn to the left and fly upriver to the north for a more suitable crashlanding site.

It was becoming unbearably hot in the cockpit before he found a flat and sandy area in the Abra River. Dropping his combat flaps to break speed and give himself a shorter landing distance he still hit the sand dune at about 110 mph. The P-38 skidded sideways before it finally came to a stop and he was able to quickly break through the overhead escape panel.

Flames were now quickly consuming the P-38 and Hammond ran for his life before the wreckage exploded. He dropped his equipment, including his parachute, on the way to the safety of some bushes and reached them just as the P-38 blew up. Taking stock, he was beginning to feel the painful burns on his arms, shoulders and face. Fortunately, he carried a small compass in his pocket and decided to take a westerly course toward the coast. It was now 2:30 in the afternoon.

He found that his goggles had been smashed flat by the impact against the gunsight, but was grateful that they probably saved him from head injuries. As it was, the pain from his burns was distracting enough in his effort to keep on the alert. He decided to hide during the day and make his way by night.

While he rested he could hear occasional Japanese patrols during the afternoon. They seemed to be passing within fifteen to twenty feet of his hiding place. Later, when he was able to make his way along a road in the diminishing light, he came upon a lone Japanese guard whom he attacked with his knife butt. The Japanese was left sprawled by the side of the road. Hammond was afraid he had killed the man, but hurried out of the vicinity before he could be sure.

Bob Weary crashlanded on a Lingayen Gulf beach on May 28, 1945 and was rescued some time later. (Krane collection)

Strikingly marked #134 with pilot other than Moering. (Krane collection)

By next morning Hammond's burns were even more painful and he treated them as well as he could. He tried following a creek that ran to the west-northwest, but nearly stumbled headlong into some Japanese camps. Then he tried following the top of a ridge leading toward the coast, but it started to rain and he retraced his steps to a large willow tree on the ridge that afforded some protection, at least.

The burns on his body troubled him and he was only partially protected by the tree, but he managed to sleep. It was the only immediate relief from the misery of his situation.

May 5 dawned bright and warm, but Hammond was not cheered by his prospects. Once again he tried the top of the ridge, from which point he at least had a better chance of seeing the Japanese before they saw him. He found another creek that led him to a broad valley which presented new obstacles.

However, there was no alternative and he descended into the depth of the thing, falling several times before he got up the other side. It was arduous enough under normal terms, but lack of food and the continuing pain of the burns took a special toll of his stamina. It was dark when he reached the other side of

the valley and his fatigue simply dictated that he lie down and sleep.

Somehow he gathered enough personal resources to get up on the morning of May 6 and continue heading west. At one point a P-38 divebombing attack on a nearby target forced him to the ground which shook under the impact of detonating bombs.

At one o'clock in the afternoon Hammond saw someone waving at him and he was overjoyed to find that it was a friendly Filipino. The native warned him of Japanese nearby and took him to a place of safety where other civilians gave him food; rice, eggs and bananas which Hammond gratefully devoured.

Later, he was taken to the camp of Lieutenant Moreno of the 121st Infantry. There he was given clean clothes and his burns were more thoroughly treated. The next day he was able to rest and began to feel better. On May 8 he endured a painful bandage change which grated on his fire-damaged nerves.

By May 10 he was on his way back to Lingayen, escorted by twelve Filipino soldiers. Ten civilians joined the procession as carriers on May 16. Three days later the young 433rd

Chris Herman by his #134. (Krane collection)

Squadron P-38 pilot was back in Lingayen. He left the Pacific for good in July and went home to Wichita, Kansas for a welcomed leave.

Ipo Dam

At the beginning of May the Japanese still had a knife at the throat of Manila since they still held the town of Ipo to the north of the capital. The town of Ipo held the reservoirs which supplied water to Manila and the Japanese had the option of cutting off the supply or even contaminating it. As it was, they shut off the water and resisted for weeks every attempt to take the Ipo dams, cutting the Manila water supply by one-third.

The 43rd Division had been freed from duty in the south because of the collapse of Japanese resistance in that direction, and was shifted to the Ipo dams area. From May 4 through May 6 American fighter-bombers dropped a quarter of a million gallons of napalm and tons of fragmentation and general purpose bombs in the Ipo area.

The 43rd Division, supported by Filipino units, made a determined drive starting on the evening of May 6. Those Japanese still resisting were seriously off-balance and allowed the Allies to take the dams intact. After May 16, the Japanese were again forced to rely on dug-in positions around the hills and forests surrounding Ipo.

Between May 17 and May 26 the 475th made some concentrated strikes on the Ipo area, using the dreaded napalm to murderous effect. The 431st Squadron was especially active during the period and scored heavily against Japanese positions.

Examples of the squadron's missions include operation number 1-901 which sent six napalm-loaded P-38s to Ipo on the morning of May 18. Lieutenant Hanway developed a runaway propeller and had to be escorted back by Lieutenant Werner. That left the four P-38s of Chris Herman, Hal Gray, Pete Madison and Jim Moering to lay their fiery charges accurately on Japanese positions.

Mission 1-925 had a larger force of fifteen 431st P-38s taking off at ten minutes before eight o'clock in the morning on May 21 to Napalm troop concentrations between the Puray and Mariquina Rivers. The pilots went in at 300 feet above the hilly terrain and devastated the Japanese positions.

Jim Barnes by his bombed up #127. General Ennis Whitehead insisted on maximum armament for every mission, thus the high proportion of 1,000 pound bombs during the Luzon ground attack missions. (Krane collection)

Mission 1-927 was a strike to the same target on May 23. Sixteen 431st P-38s took off at 7:35 in the morning and began orbiting over the target at 9:45. Fifteen P-38s went in dropping their bombs, Lieutenant Ray North having turned back with low fuel pressure.

The results were termed excellent for this mission as well as the series of strikes on the Ipo area in general. When the 43rd Division swept through the valley and captured the Wawa dams intact as well, its commander sent the 475th a commendation for the ease with which his unit swept through the area.

Into the Cagayan Valley

With resistance in the south crumbling rapidly, the 475th was assigned to strikes in northern Luzon as well as missions to the Asian coast. The 432nd flew two missions to the Balete Pass area on June 3 with excellent results. On the first mission, eight P-38s took off at nine a.m. to bomb enemy concentra-

tions northwest of Balete Pass. After the bombs were dropped and twenty-four strafing passes were completed the controller reported ten to twelve supply buildings set afire on the east bank of the Pambang River.

The second mission had seven P-38s getting off at 10:30 to hit a wooded patch of ground five miles southeast of the Pass. Again the controller was well pleased with the bombing and strafing.

Bombing of the Balete Pass continued intermittently with long range strikes for the 432nd through June 6 for the 432nd. On that date two more missions were flown north of the Balete Pass. The first set fire to more than a dozen buildings and another, probably used as a warehouse, was riddled by strafing.

The second was even more eventful when eight more P-38s took off in the afternoon to bomb a bridge behind the retreating enemy. On the way in, the P-38 of Lieutenant Arthur Goodwin caught fire in its left engine and he was forced to bail out nine miles northeast of Dagupan. He landed safely and was picked up by friendly troops immediately.

Another P-38 aborted the mission with mechanical trouble, leaving six P-38s to finish the job. Those six dropped their bombs and scored three direct hits on a steel bridge and two more on a wooden bridge slightly to the north. The remaining seven bombs landed around the approaches to the wooden bridge and started a fire.

After the Americans had forced their way into the Cagayan Valley by the beginning of June, Yamashita moved his headquarters closer to the Bontoc area. The 32nd Division had penetrated the Salasac pass and joined the 25th Division north of the Balete Pass. Yamashita still had hope of counterattacking the Americans, but his enemy was steadily advancing into the valley and by June 26 American paratroops landed in the north near Appari and were driving south.

On June 10, the converging American and Filipino ground forces reached Bambang at the eastern base of the formidable Baguio-Bambang-Bontoc triangle. The American 6th, 32nd and 33rd Divisions were pressing north to Bontoc through difficult mountain terrain while the 37th Division moved rapidly into the Cagayan Valley.

The Japanese were trying to live off the land, forming into small guerilla units with the limited goal of making Americans move slowly on dangerous ground. Luzon was quickly becoming secured and by the middle of the month the 475th missions were generally being pared down for lack of specific targets.

On June 13, the Satan's Angels were busy depleting the Japanese Army and depriving it of its meager stores in a general attack on the Camalaniugan area, just south of Aparri on the north Luzon coast. The area was crowded with a confusion of Japanese men and materiel that received a terrible pounding from strafing and 1,000 pound bombs. Later

Bob Maxwell and H. Mayer inspect a Zero – probably of the 121st Naval Air Group – on Clark Field in June of 1945. (Maxwell)

475th attacks could not be accurately assessed because of the smoke from previous bombings.

The 432nd Squadron flew four missions on June 15, two divebombing and two strafing. On the first the P-38s divebombed Japanese troops retreating on foot along Route #4 and made eighty strafing passes that left at least fifty Japanese dead along the roadside.

The last 432nd mission of the day was a general strafing sweep along the same part of Route #4. Five P-38s took off a few minutes after two o'clock in the afternoon and ranged over the target area for an hour and forty-five minutes. Ten to fifteen Japanese soldiers on foot were killed in the sixty strafing passes.

On June 18, the 475th began its first long range napalm missions with a fire bomb raid to Kari, a town north of Tainan in the southwestern part of Formosa (Taiwan). The entire town was left in flames by the terrible efficiency of the Napalm.

Pappy Cline near the end of his tenure as commander of the 431st Squadron (Krane collection)

Bob Werner by #135 when it was taken over late in June by Major Vogel, the new commander of the 431st, and renamed MISS MANOOKIE. (Krane collection)

475th boneyard in the Philippines. (Scott Ferris)

It was not without cost, however, and the 431st paid the price for bombing success. One of the friendliest of the squadron pilots, according to the 431st official history, was Lieutenant Alvin G. Roth. His P-38 was likely hit by small arms fire in the right wing tank and he was last seen trying to get out of his cockpit at a very low altitude about five miles southwest of the mouth of the Subun-Kei River.

Squadron newcomer Lieutenant Edward Carley was apparently also hit by small arms fire at very low altitude. He tried to bail out, but subsequent searches by rescue aircraft failed to locate either pilot.

Another unfortunate 431st casualty happened on a June 21 mission to Tuguegarao, almost due south of Aparri. Lieutenant Herb Finney was a Satan's Angel who was also known as an avid basketball player. He had been flying missions for some months when he was scheduled for the bombing mission. Just after he took off, white smoke was seen trailing behind his P-38 and he apparently abandoned the fighter too low to survive.

Distinguished Unit Citations One and Two

The day before the 431st Squadron lost Finney it was gathered on Lingayen to receive the first two of the DUCs it would receive in World War II. General Kenney commended the group by saying that the "achievements are keeping with the finest traditions of the Armed Forces of the United States" and "reflect great credit on the personnel of the 475th Group and the United States Army Air Forces."

The first DUC cited the outstanding performance of the group in action on August 18 and 21, 1943. In the language of the citation: "On both days the P-38's of this group had to battle large numbers of intercepting fighters in fierce engagements carried on over a large area. Maintaining formation throughout the entire combat in spite attacks from every direction, the pilots of the 475th Fighter Group not only successfully defended themselves and protected our bombers, but also definitely destroyed 38 enemy planes and probably 6 others."

This citation also gave some well-deserved credit to the groundcrews: "Credit for the achievements of this group must be given not only to the P-38 pilots who fought so skillfully and gallantly but also to the ground personnel who so effectively prepared the aircraft for these missions. The accomplishments of the 475th Fighter Group are in keeping with the finest traditions of the Armed Forces of the United States."

The second citation was for the actions in Oro Bay on October 15 and 17, 1943. For the October 15 mission the action of the 475th was worded thusly: ". . . the Lightnings fought in flights and the enemy planes scattered, fighting singly and in pairs, making it difficult for our fighters to concentrate their fire power. However, in a series of spectacular actions, the P-38s destroyed all the divebombers and allowed not more than 7 enemy fighters to escape... the 475th Fighter Group prevented a serious loss to vitally needed American shipping."

Wording for the October 17 mission was no less laudatory: "2 squadrons of this group climbed to meet approximately 30 enemy fighters approaching Oro Bay at altitudes of 22,000 to 24,000 feet. Demonstrating superb teamwork and exceptional flying skill, the pilots of the 475th Fighter Group again pressed fierce attacks against the enemy. In rapid succession the Japanese aircraft exploded or crashed into the sea in flames and the hostile formation was scattered. . . . the total for the 2 days of combat (was) 56 enemy aircraft destroyed and 8 others probably destroyed, a remarkable record for 2 fighter engagements."

The 475th Group: was honored with yet a third Distinguished Unit Citation that was presented after the war. This last citation honored the divebombing and interception activities of the group during the period October through December 1944 in the Philippines.

General Kenney personally travelled to Paterson, New Jersey to present the Medal of Honor posthumously to Major Thomas B. McGuire, Jr. on May 8, 1946. If nothing else is known about McGuire, he was an absolutely fearless fighter pilot who brought one last distinctive honor to the 475th Fighter Group's World War II record.

After June 26 there was little left to capture on Luzon. The paratroopers and ground forces of the American I Corps linked up at Tuguegarao, leaving the Japanese with nothing more than several large pockets of resistance.

Missions for the 475th were in decline throughout the rest of June. One diarist boasted that the pilots were enjoying "banker's hours", sometimes flying one mission a day that took off in the afternoon and ended less than an hour later.

On June 30, for example, the 475th flew fewer than fifty sorties, dropping napalm and high explosives on Japanese positions in the ragged lines near Bontoc.

More Rotations

When General MacArthur declared Luzon secured on June 26 he also tacitly approved the new point rotation system. The older system simply had the men with most time in the theater scheduled to be sent home on an intermittent basis. If the conditions were adverse to releasing troops for rotation they could be stuck in place for the duration.

Since the war in Europe had ended and many new troops were being trained in the U.S., there was every chance that any one position would be filled many times over. Hence, if someone could accumulate at least eighty-five points under the new system, he would automatically be ordered home.

If an enlisted man would, say, have spent time in the hospital, had some exposure to combat missions, won an award for heroism or efficiency and have no blots on his record, he would be high on the list for rotation if he had spent a minimum of one year in the theater.

The 475th was returning dozens of enlisted men to the U.S. by the middle of 1945, while the average was no more than four or five a month before that time. About forty-five enlisted men returned home in June, and the rest were jubilant in their chances of seeing American shores very soon.

One of the pilots who got his orders in June was Captain Perry Dahl, who left for the 262nd Replacement CO., 93rd Replacement Battalion at Leyte on June 6. He had been aptly named "Lucky" in at least three narrow escapes on operational flights. Many pilots from the group had not escaped from similar situations.

Dahl was the 432nd Squadron Operations Officer at the time he left the squadron. He had flown 158 combat missions and was officially credited with nine Japanese aircraft shot down. At the beginning of his tour he was one of the youngest fighter pilots in combat and by the end he had given his country sterling service.

The end of the war was coming quickly for the 475th Fighter Group. If it had been born quickly in the fires of the war, then it would subside just as quickly in the coming time of peace.

Replacement pilots of the 432nd Squadron in August of 1945. Background is the bustling airfield at Ie Shima. Each of the pilots was well-trained with at least 175 hours in P-38s. Bob Maxwell is second from left. The others are, coincidentally, pilots with names that start with M; Mayer, May and Mills. (Bob Maxwell)

Chapter XVI

ON JAPAN'S DOORSTEP JULY-AUGUST 1945

By the first week of July the rotation system had speeded up. More than forty enlisted men were scheduled for a stateside return before the end of July. Pilots going home during the month included Bill Richmond, Bert Roberts, LeRoy Ross and Ed Edesberg of the 433rd. Dean Olson, Harrold Owen and Robert Koles of the 432nd also departed between July 16 and 21.

Major Elliot Summer was hospitalized with appendicitis and Major Dean Dutrack took over command of the 432nd on July 5. "Pappy" Cline turned over command of the 431st on June 23 to Major John Vogel who in turn surrendered command to Ed Weaver on July 16. Cal "Bud" Wire had gone home in May and Captain James Wilson took over the 433rd. Squadron records were signed by Captain Bill Haning in July, suggesting that Wilson was rotated home and Haning took temporary charge of the squadron.

In effect, the last of the combat veterans were being sent home while newer faces took on the burdensome task of finishing the war and preparing for occupation duty. These latter day Satan's Angels carried on in true 475th Group standard.

One of the original members of the 475th to leave in July was Group Sgt. Major MSgt Clay V. Cockerill. He had been with the group since Amberly Field, but was rotated home under provisions of the "over 40 years of age" ruling.

Satan's Angels on Top

Weather hampered 475th operations for the first few days of July. Not that it mattered much, since all major military objectives had been accomplished and the pace of ground support operations was now decidedly selective. For example, one mission that was scheduled for July 9 was canceled because the controller's observation L-5 developed some mechanical difficulty and he returned early.

That same day Joe Forster returned from a ground gunnery school course at Foster Field, Texas. He had been gone since March 24 and was somewhat surprised to find the 475th in relative inactivity.

But the pace satisfied most of the crews, especially the groundcrews who had some time behind them. It was pleasant to have lots of time for maintenance rather than the rush of intensive combat operations. Some days were turned over completely to maintenance with nothing more than training flights scheduled.

The tacit and sometimes spirited rivalry between the 49th and 475th Fighter Groups had also mellowed during the first weeks of July. Both groups had scored their 500th confirmed aerial

First Lieutenant George Smith beside his #130 MISS GEE GEE. He was a 431st Squadron pilot, but is listed as Captain George W. Smith, Jr. from Headquarters, flying as element leader in a 431st training flight when he became the last known 475th casualty on July 11, 1945. (Krane collection)

Nose of MISS GEE GEE sometime in early 1945. (Krane collection)

victory during the Leyte campaign and continued trying to best each other even after there were no more Japanese in the skies.

It was too late in the game for the 475th to top the 49th's record score of well over 600 aerial victories, but the Satan's Angels could boast of thirty Japanese aircraft caught and destroyed on the ground during February and March 1945, and of a bag of ten locomotives destroyed on Formosa during April. Added to countless Japanese soldiers killed and vehicles or equipment destroyed, the 475th could stand on its own record in the Philippine campaign.

One of the missions that added to the record of the 475th was the last of the day flown by the 432nd on July 15. Four P-38s took off at three o'clock in the afternoon to strafe trucks in the diminishing Bontoc triangle area. Within twenty minutes the P-38s were over the target area and made thirty-two strafing passes. Four trucks were listed as destroyed and the 432nd fighters were back on the ground in little more than an hour after they had taken off.

On July 14, Brigadier General Freddie Smith, commander of V Fighter Command, came down to Lingayen Gulf and awarded 151 decorations to 75 Satan's Angels pilots. Those assembled on the white beach of Lingayen witnessed the last wartime presentation of awards to 475th personnel.

The Move to Ie Shima

Both the 49th and 475th Fighter Groups were relieved of combat duty on July 23, 1945, in preparation for the move to the Ryukyus. The 431st flew a typical mission, number 1-

1006 on July 20, when seven of its P-38s went out on a firebombing and strafing strike. The next day it flew mission number 1 1008, its last in World War II. Another firebombing and napalm mission was number 2-907, the last for the 432nd. The 433rd flew a convoy cover over the Luzon Straits on July 21 for its final operation, mission number 3-1047.

The Cagayan Valley would be home to a receding series of hostile Japanese troop pockets for the remainder of the war. There was little need for first-line air attacks on these battered, sick and starving enemy troops. War in the Pacific was now concentrated on the Japanese homeland itself.

On July 26 the three squadrons and Headquarters tore down all the structures that had been their camp and loaded trucks that headed for White Beach. The 431st loaded aboard LST 793, the 432nd aboard LST 752 and the 433rd aboard LST 1014. By 9:30 the next morning all loading was completed and the ships pulled out of Lingayen Gulf, headed for Subic Bay.

Most of the personnel of the three squadrons, with Headquarters accommodated aboard LST 752, entered Subic Bay on July 28 and remained there until the last day of the month. A few of the crews were trailing the LSTs in smaller and even more cramped LSMs. On July 31 the 475th Group set sail with about sixty other ships for what had been revealed as the ultimate destination: Ie Shima. What the men of the 475th would eventually learn was that the island was little more than a rock sticking out of the ocean a few miles from Okinawa. The top of the rock had been hacked flat for the use of airstrips by hundreds of warplanes. There was little else on the island besides whatever was needed for the operation of those planes.

Japanese surrender Betty on Ie Shima, August 19, 1945. (Hanks)

475th personnel, among others, watching the white painted Bettys with their green surrender crosses landing on Ie Shima. (Haines via Maxwell)

For the moment, the 475th was enjoying exceptionally good Navy food, including fresh meat and vegetables, and relaxing aboard the ships. This bit of unusual luxury for the Satan's Angels lasted until August 6 when the convoy dropped anchor in Okinawa's Yonabaru Harbor.

Late in the night of August 6/7 the 475th experienced its first air raid alert since the Clark Field alert on March 5. The two hour alert was filled with the droning of Japanese aircraft and what was taken to be bombs exploding, but nothing came near enough to threaten the ships carrying the Satan's Angels.

Last of the War

Two more air raids were endured on the night of August 7/8, but no damage was done and the Satan's Angels were too excited to be distracted by mere bombs falling. The announcement of the atomic bombing of Hiroshima both stunned the men and filled them with anticipation. Perhaps the end of the war could be closer than anybody thought?

Then within forty-eight hours two more hammer blows came. First, the perfidious Stalin, who since December had been interceding on the behalf of the Japanese with the Allies, declared war on the people for whom he had been transmitting peace terms. Then the second atomic bomb was dropped on Nagasaki.

One of the factors that was prolonging the war was the manifest American decision to invade Japan. The Japanese reasoning was that an invading American Army would sustain such casualties that more liberal peace terms would be possible. Such an approach would not be so conducive to a favorable conclusion with the less humanitarian concerned Russians who immediately began invading Japanese territory.

Whatever the intentions of the various warring nations, events were obviously developing their own momentum with whirling speed. On August 10, the day that the 475th water echelon pulled out of Yonabaru Harbor and the air echelon flew to Ie Shima, the Japanese government asked the Swiss and Swedish governments to transmit its acceptance of Allied peace terms.

All 475th personnel were encamped on Ie Shima by the evening of August 12. Major Dean Dutrack of the 432nd Squadron had been going about the chaotic work of packing up the squadron and moving it to Ie Shima. Elliot Summer had come back to the squadron on July 17 and was in charge until he was rotated back home on July 28 just in time for Dutrack to take permanent charge of the 432nd at the most hectic moment of the move.

In spite of the overwhelming dizziness of the details involved, Dutrack presided over a successful move to the Ryukyus. He stayed with the 432nd and 475th in other capacities until 1947.

The Satan's Angels lived a makeshift life bathing out of their helmets, for one thing, until word was received that V-J day was proclaimed on August 15 (August 14 east of the international dateline) pending formal surrender ceremonies to be held in the weeks ahead. Even so, that same night a Japanese aircraft dropped two defiant bombs on Okinawa and 475th crews followed the chase over the radio: of F4U Corsairs trying to pin down the elusive little Japanese gnat without success.

On August 19 the two Betty bombers in their white surrender finish with the green crosses where angry red hinomaru national markings had formerly been landed at Mocha strip on Ie Shima around noontime. The Japanese delegation, headed by Lt. General Tatashiro Kawabe of the

Bob Werth's #133, pickled for movement to either Ie Shima or Korea. 475th records indicate that IRISH ANGEL did not survive until December 1945 on the group's roster.

Imperial General Staff, glumly stepped down from the bombers.

An American reception committee led by General Freddie Smith curtly greeted the Japanese. An hour later the even more rigid faced members of the Japanese delegation boarded a C-54 transport plane and were off for Manila by 1:30 in the afternoon. The Betty bombers were moved to an adjoining taxiway to await the return of the peace delegation which would not be back until the last details of peace were negotiated.

Bill Hasty's Odyssey Part II

About the same time that the Satan's Angels were looking over the Japanese surrender Bettys and the Japanese peace delegation, Bill Hasty was being informed by the first free American he had seen in over a year that he and his fellow prisoners would be repatriated within the next few days. It was hard to believe that all the privation and abuse would be over forever.

Hasty had arrived at the Japanese Naval camp around the fourteenth of June, 1944 and was immediately placed in solitary confinement. For the next three months he was brutally interrogated concerning the range and fuel capacity of the new P-38s. During that period of time he was threatened and beaten around the clock at two-hour intervals.

It was ironic that the Japanese had spared his life to get information that Hasty simply did not have. He was familiar with the P-38H model and its external tank system, but could tell them little or nothing about the newer J model with its outer wing leading edge tanks. The English-speaking Japanese interrogators simply did not believe him and continued the intensive grilling.

During the intervals between questioning, Hasty would lie on a bare floor in a 4'x6' concrete cell. He was given a rice ball and boiled water twice a day; once in the morning and once at night. There was no contact with any other human being except the interrogator.

One day in September the ragged-looking American prisoner was let out of his dungeon. His hair and beard had grown to wild proportions and he was much thinner than his normal 165 pounds.

The other inmates greeted him with enthusiasm since he could perhaps give them news of the D-Day invasion in Europe. Bits and pieces of news came from snippets of overheard conversations of the guards or gleaned from whatever could be translated from discarded Japanese newspapers. Since Hasty had been shot down exactly one day before the invasion, he could not have known anything about it and was a subsequent disappointment to his fellow prisoners.

One of the first people to greet Hasty when he left his tiny cell was none other than Greg "Pappy" Boyington, the famous Marine Corps ace who was shot down in an F4U Corsair the preceding January. Boyington had just shot down three Japanese fighters to unofficially break Rickenbacker's record with a tally of twenty-four Japanese in the air and four others on the ground. The Japanese fished him out of the George Channel between New Britain and New Ireland and treated him with mixed contempt for his defiant attitude and admiration for his combat record.

Hasty learned that Boyington stood up to the same sort of third degree that he had endured and, in fact, became a legend in the camp for not giving in to the Japanese. As a result Boyington had "got the hell beat out of him on numerous occasions with 5 ft bamboo poles" every time he boldly defied the guards.

Around the end of September or beginning of October Hasty was in a group of prisoners that was moved to the HQ Tokyo camp. The senior American officer in the camp was a Commander Mayer, formerly the gunnery officer aboard the USS Houston, the cruiser sunk at the battle of the Java Sea early in 1942. The commander was one of many officers that Hasty came to admire in the difficult days of captivity.

The officer prisoners were made to work in the compound leather shop while the enlisted men were forced to do standard labor outside. At least life took on some of the rough outline of routine with a bit too little food and comfort and a bit too much insult and brutality.

On November 1, 1944 Hasty caught sight of the first B-29 reconnaissance mission over Tokyo. He was impressed by the huge vapor trail of the new American bomber flying high above in the thin atmosphere. It was cheering to see the smaller contrails of the Japanese fighters vainly trying to catch the bomber that flew so swiftly in the higher reaches.

Sometime later another B-29 witnessed by the prisoners in the compound was not so lucky. It was chased right down to the water by a number of Japanese fighters and crashed into Tokyo Bay with the enemy at its heels like a pack of hounds.

In February Hasty and the other prisoners were treated to the spectacle of the first U.S. Naval air strike on Japan since the combined Army-Navy Doolittle Raid of April 1942.

The end of the empire and some of the 475th men who helped reverse the rising sun.

Above and opposite: Unidentified young 431st pilot who came into the squadron late in the war and probably went with the group on occupation duty in Korea. (Krane collection)

Scores of Navy fighters and bombers hit airfields, docks and other targets around Tokyo Bay during the day on February 16.

Those American prisoners who saw the divebombers begin bomb runs right over their heads were certain that the pilots were aware of the compound and did what they could to both avoid hitting it and encourage the captives. It was a heartening moment that encouraged these threatened Yanks to believe that the homeland was at least aware of their existence.

Late in February Hasty was in another group of prisoners that was moved to Yokohama's "Brickyard camp." The Japanese may have perceived the actions of the American Navy planes the same way as the prisoners and decided to scatter their captives.

On May 14 the first four wing B-29 raid hit the Yokohama area and left blackened earth around the brickyard camp. A few stray incendiary bits hit the camp, but the resultant fires were small and easily extinguished.

Hasty watched B-29s shot down on several earlier raids. He was puzzled that the B-29s seemed to come over in trail at the same altitude, making the work of the Japanese antiaircraft much simpler. Some of the bombers were sitting ducks in such a situation and Hasty was horrified to see some of the crews bail out with parachutes on fire, collapsing after a few moments and dropping the crewmen to their deaths.

A few days after the May 14 air raid Hasty was moved again to a camp in the town of Niiagata, just north of Tokyo. There were about eighty men and six officers finally gathered together in this camp, many of whom had amoebic dysentery among other serious afflictions. The daily diet was boiled soy beans and barley with occasional fish heads added in.

Sometime late in July or early August the Americans began to notice a remarkable change in the attitude of the guards. What had formerly been an attitude of cruel contempt had changed to a palpable deference. Around the thirteenth of August the Japanese camp commander called the six prisoner officers and curtly told them that he was turning the control of

the place over to them as he had urgent business in Tokyo!

The six Americans called a meeting of the prisoners and told them the news which was received with thunderous cheering, even from the sick and exhausted men. On September 5 the Americans commandeered a train and went to Tokyo when the promised relief did not arrive at the camp. Some U.S. Navy planes had dropped cigarettes and candy a couple of days after the Japanese surrendered and B-29s parachuted food in 55 gallon drums, but these captive Americans were ready for liberation.

Upon arriving in Tokyo the Americans were met by General Krueger, whose Sixth Army had been supported by the 475th in liberating the Philippines. Hasty and the rest of the former American prisoners were deloused, cleaned up and issued new khaki uniforms. Medical treatment was also provided for those who needed it.

Hasty and the other prisoners who could walk were sent to Atsugi airfield and loaded aboard C-54s. They flew to staging areas in Okinawa and at one point were greeted by none other than General Jimmy Doolittle who personally welcomed each man home.

Two days later Hasty was in Manila for POW debriefing. He was able to send a telegram home saying that he would arrive around the middle of October. He and five thousand other former POWs were loaded aboard the Hugh Roddman and set sail for the Golden Gate. He was able to eat steak and eggs or ice cream at two hour intervals or anytime he wanted them.

Hasty was the only 475th prisoner of war to make it back home. There were others who succumbed to the brutality of the wartime Japanese military. They were part of the payment for the success of the Satan's Angels.

Lt. Roger Palmer's P-38 with Bob Maxwell in goat-fur boots during the cold winter (February 1946) at Kimpo. (Maxwell)

Chapter XVII

SATAN'S ANGELS POSTSCRIPT

T he 475th Fighter Group did a great deal of firebombing during its last months of combat. Japanese reaction to the group if any of that reaction could have been recorded must have been that these P-38s with the horrific demon or lightning motif were indeed from the infernal regions.

During the final five months on Luzon these were the tonnage figures for 475th Group delivery of high explosive and napalm:

High explosive	Napalm	
82 tons		Feb. 1945
901	284	Apr
889	620	May
1195	576	June
810	103	Jul

From the beginning the 475th was destined to be a source of difficulty for the Japanese. If the group was not aggressively attacking the enemy in the air it was hurting him without mercy on the ground. The tradition of the Satan's Angels was thoroughgoing efficiency as a combat unit, and its members paid their dues to be a part of it.

Sgt Vince Steffanic had been rated a top communications chief in the 36th Fighter Squadron when he found himself at Amberly Field on June 17, 1943 in the new 432nd Fighter Squadron. His first job was to interview each incoming man to determine various qualifications. Using a homemade shipping box cum office desk to both get a sense of each man and to orient them into the high standards expected in the new organization, he got quick results and cooperation.

Any man who did not always reach the high standards of the 432nd Squadron radio section would be assigned an unappealing task such as the monotonous switchboard as a sort of discipline. Thus, even in the less combat-related jobs, the 475th fostered excellence. Part of the discipline for pilots included removal from combat for breaking the strict rules. The 475th was a hard group to get into and an easy one in which to receive harsh discipline.

At the end of August 1945 the occupation troops of the Satan's Angels enjoyed a more relaxed atmosphere out of the stress of wartime conditions, but began to flinch under a new set of restrictions that applied to peacetime. More spit and polish and less of the shock troop roughness that Charles Lindbergh had observed became the order of the day.

By August 29, 1945 the move that would be the 475th Fighter Group's first postwar station was announced. One of the final entries in the 432nd Squadron's operational diary comments: "

Unloading spare P-38 engines for the 475th September/October 1945 at Inchon. The rapid tide ebb – six feet per nour – gave only a one-hour window for unloading. Sea wall at right was Red Beach during the October 1950 Inchon landings of the Korean War. (Maxwell)

. . our future home would be Korea – just what particular section of the peninsula we are going to is not yet known. . . . we are still in the dark as to exactly what type of country we are going to and what sort of people and customs we will encounter the boys who expect to be in the Army of Occupation are hoping for the best."

The new base would be Kimpo, Korea, near Seoul. One of the first big tasks that Col. Loisel undertook as commander of the 475th was the movement to Korea and the job was done by the last week in September. It was the fate of American units in Korea to be relegated to the backwater of occupation duty and the 475th suffered the usual deprivation of such units.

Inspection at Kimpo, probably in the autumn of 1945. Reviewers are passing 431st Squadron P-38 #118 and receiving salute from crew of #138 MOMMY.

Probably the same inspection. Crew in foreground stands ready with Elliot Summer's old BLOOD AND GUTS.

Inspection at the 433rd line with shining new #198 and #177. 177, serial 44-26473, lasted until at least the first part of December 1945.

One of the things that the group had never faced during the Pacific war was a cold and bitter winter. The first endured at Kimpo was difficult indeed, and both men and P-38s suffered as they hadn't done in the heat of the Southwest Pacific. The lack of both spare parts and warm clothes was an enemy just as formidable as the Japanese had been.

January 1, 1946 dawned gray and chill at Kimpo. All activity was especially sluggish because of the preceding night's party. Friction with the Korean population prohibited the invitation of female guests, adding to the presence of strong drink as the primary activity of the evening.

On the first day of the year there were 116 officers and 128 enlisted men in the 475th. By the end of the month there were only 97 officers and 63 enlisted. The desire to be anyplace but cold Korea encouraged many to go home at the end of their tour and the reputation of the place discouraged many others from volunteering for service there.

It was a time of general reduction in forces. Nobody seemed to want peacetime military service with its niggling discipline and bleak prospect for the future. Everyone seemed to want the military cut in order to get on with peace and prosperity. In effect, those units in places like Korea had become somewhat like frontier outposts, understaffed, underequipped and largely forgotten.

Cold weather hampered activities at Kimpo. In an improvised attempt to improve working conditions Japanese flying clothes were issued. At the end of January a central heating unit was installed in the Hangar, but lack of supplies and parts deferred its use.

In addition, the P-38s were having their own problems with the cold. The 475th hadn't come across the factor of low temperature before, and their P-38s experienced the same

effect of congealing oil that had plagued crews in England. Oil dilution was not being satisfactorily accomplished on P-38s with the result that many of the aircraft were off operations.

Of sixty-nine P-38s assigned in January only forty eight were flyable. The end of January and beginning of February found the 475th at its ebb in both men and aircraft. As a consequence, all three squadrons were barely operational, the 433rd Squadron, in fact, was completely off operations until June. The plan to bring the group up to operational strength had the 432nd scheduled to be fully capable by May 1 and the 431st by May 15. The 433rd had a longer way to go since it was off operations completely, so wasn't scheduled for full strength until June 1, 1946.

In order to achieve full pilot staffing, some P-51 pilots were transitioned to the P-38. Most of their comments were non-commital, but some of them ranged from, "Its a fine ship" to "Well, here we are (fighter pilots flying bombers)."

In April 1946 Lt. Colonel John S. Loisel was succeeded as commander of the 475th by Colonel Henry Thorn. Loisel had been overseas since early in 1942. He had worked his way into the hornet's nest of combat when he joined the 36th Fighter Squadron in July 1942 and the 432nd in May 1943. He flew 301 combat missions and was reputed to have been in the Southwest pacific and Far East longer than any other pilot or commander of the period.

The job that the 475th was assigned consisted of patrols over western Korea to the northeast corner of China. Throughout 1946 the P-38s of the group were meant to demonstrate the military strength of the west as well as reconnoiter possible armed threats. In addition, the group participated in ground and air displays to promote goodwill with its South Korean host.

432nd P-38s (nos. 150 and 152) guarding the remains of 85th Sentai Ki-44s and Ki-84s at Kimpo.

During this time the P-38 was fading from sight and the P-51 was taking its place. In April it was decided to reactivate the 433rd as a checkout squadron for P-51s and thus all P-51 pilots and enough P-51 aircraft were assigned to the squadron thereafter.

One of the arguments for conversion to P-51s was the fact that all P-38s in military use were showing signs of strain. Many of even the latest P-38s were developing stress fatigue around the wing and boom areas. In April there were two fatal crashes that perhaps could be related to stress problems.

P-51 D and K model flying time began to appear on 475th monthly reports for June 1946. Most of the P-51 time was probably generated in the 433rd throughout August and September until the number of P-51 missions began to exceed P-38 missions in October. By December all missions were flown in P-51s.

Throughout 1947 the P-51s of the group continued the routine of patrol and occasional aerial displays. By 1948, however, it was considered prudent to consolidate American air strength in Japan. The need for an air force right on the cutting edge of conflict with the Soviet Union was riskier than the benefits it offered. If air cover were needed in the likely event of conflict between North and South Korea, then bases in Japan could respond almost as quickly and just as efficiently with less chance of being overrun.

By the end of August the 475th and its force of P-51s had moved to Itazuke, Japan. Even this move did not serve to halt the group's decline from the crack fighter unit it was during the war to a burden on the policies of the day. With declining effectiveness in the light of emerging jet fighter technology, the P-51 equipped units were obliged to consolidate and reduce the scope of their usefulness.

Some units reequipped with F-80 – and later with F-84 jets during the Korean War. Some of the P-51s were still useable right through the first stages of the conflict and were gathered into a few squadrons.

The 475th was selected to facilitate the reorganization of the brand-new Far East Air Forces by being disbanded. The Satan's Angels made one last move to Ashiya in March 1949, and was inactivated on April 1. It seemed as though the group simply slept through the Korean War. Those P-38s with which it had survived World War II and the immediate months afterward had been cut up for scrap, but the heart of the 475th was still alive. In September 1955 the 475th Fighter Group (Air Defense) was reactivated and revitalized at the Minneapolis-St. Paul International Airport.

195 and 188 on the line at Kimpo Airbase, 1945.

Ken Holdren, Frank Nichols and "Red" Herman, all formerly of the 431st Fighter Squadron, at a 475th reunion during the mid-1960s. (475th Fighter Group Association)

The author can remember as a young teenager seeing the bright silver and red F-89 Scorpions flying overhead in the skies of Minneapolis. He even got close to some of the machines on special days when the F-89s were on public display without ever realizing the full impact of the 475th's noble history. It took years of tracing the group's story to come right back to the experiences of my youth.

The 475th operated in Minneapolis minus the 431st Squadron, which had been activated in 1953 with the 7272 Flying Training Wing. Operating F-86D radar-equipped Sabres from Wheelus Airbase, Libya, the 431st provided air defense for that portion of the North African Mediterranean coast.

In 1958 the 431st Fighter Interceptor Squadron was moved to Zaragoza, Spain. Early in 1960 the squadron began conversion to the F-102 Delta Dart and was still using them about the time that it was transferred to Southeast Asia in the mid-1960s.

Meanwhile, the 433rd had converted to the F-4 Phantom and was assigned to the 8th Tactical Fighter Wing. Often led by veteran ace, Colonel Robin Olds, the 433rd FIS scored a number of victories over MiG fighters, raising its historical total to over 130 enemy aircraft shot down.

By this time the 432nd Squadron seems to have disappeared entirely. Its last association with the 475th ended with the other squadrons in 1949 and nothing more is recorded. The 432nd Tactical Recon Wing was active in Viet Nam, but no firm association with the valiant Satan's Angels Squadron can be established. This sometime rock center of 475th Group World War II operations is conspicuous by its absence.

But the other two Squadrons live on. The 431st is still operating at George AFB in Victorville. At least well into the 1980s the 433rd was operating F-15 Eagle fighters at Nellis AFB, Nevada. With defense cuts promising to end decades of cheerless dependence on costly military might it is at least possible that the last remaining units of the old 475th Fighter Group shall simply fade away.

Charles Lindbergh had the most brittle comments about the nature of the 475th Fighter Group. They were salty shock troops who gave enemy aircrews little quarter and no comfort at all. Their gaudy and lustily-painted P-38s were the scourge of the Japanese from Dobodura to the Asian mainland. The Japanese called them the "Bloody Butchers of Rabaul", and the Satan's Angels relished the title.

Perhaps someday the human race may gather enough good sense to finally abolish war. Even so, it is good to reflect that such capable warriors championed our struggle in World War II.

APPENDIXES

475th Fighter Group Wartime Commanders

Group

Lt. Col. George W. Prentice	13 May 1943 – 26 Nov. 1943
Col. Charles H. MacDonald	26 Nov 1943 – 4 Aug 1944
Lt. Col. Meryl M. Smith	4 Aug 1944 – 12 Oct 1944
Col. Charles H. MacDonald	13 Oct 1944 – 14 Jul 1945
Lt. Col. John S. Loisel	14 Jul 1945 – 18 Apr 1946

431st Fighter Squadron

Maj. Franklin A. Nichols	1 Jul – 19 Nov 1943
Maj. Verl E. Jett	20 Nov 1943 – Apr 1944
Maj. Thomas B. McGuire, Jr.	28 Apr 1944 – 23 Dec 1944
Capt. Robert F. Cline	23 Dec 1944 23 June 1945
Maj. John Vogel	23 Jun 1945 – 16 Jul 1945
Maj. Edwin Weaver	16 Jul 1945 ?

432nd Fighter Squadron

Maj. Frank D. Tomkins	June 1943 – 16 Jan 1944
(Capt. James Ince – Interim 16 – 22 January 1944)	
Maj. John S. Loisel	22 Jan – 4 Aug 1944
Capt. Henry L. Condon II	4 Aug 1944 – 2 Jan 1945
Capt. Elliot Summer	2 Jan – 5 Jul 1945 (also interim 17-28 July)
Maj. Dean Dutrack	5 - 17 Jul 1945,
	28 Jul 1945 – 1947

433rd Fighter Squadron

Maj. Martin L. Low	May 3 – Oct 1943
Capt. Daniel Roberts	3 Oct 43 – 9 Nov 43
Maj. Warren R. Lewis	9 Nov 43 – Aug 44
Capt. Campbell P. M. Wilson	Aug 44 – Dec 1944
Capt. Calvin Wire	Dec 44 – May 1945
Maj. James Wilson	May-July 1945
Capt. William Haning	July 1945? – ?

Appendix B

475th Fighter Group Losses

Pre-Combat Training Losses

July 5, 1943 Probably HQ Squadron. 2Lt Richard E. Dotson, Jr. No other mention is made of this loss other than Lt. Dotson was killed in a training accident.

August 9, 1943 Probably P-38H (431). 2Lt Andrew K. Duke was listed as killed in a flying accident.

Combat-Related Losses

August 18, 1943 P-38H1, 42-66572 (431). 2Lt Ralph E. Schmidt was the first operational fatality in the group when his P-38 disappeared during the low level cover of B-25s over Wewak. He left the formation somewhere in the combat area and was not observed to be pursued by Japanese fighters. There is no apparent link-up with known Japanese claims of P-38s shot down. He was from Los Angeles.

August 20, 1943 P-38H, (432). Lt. Allen Camp was in Capt. Tompkins flight during a bomber escort to the Wewak area. A few Japanese fighters were sighted but no combat resulted. Camp landed at Marilinan to refuel and apparently became disoriented during the flight back to base. He overshot the strip and crashed in the Karema River inlet near Terapo Mission. He was from New Jersey.

September 4, 1943 P-38H5, 42-66748 (432). Lt. Richard Ryrholm was in a flight led by Colonel Prentice that engaged a pair of Japanese fighters over the Lae area. Prentice claimed a Zero over the Hopoi area at 0830, but Ryrholm's P-38 was seen falling in flames and he was subsequently listed as MIA. He was from southern California.

September 8, 1943 P-38H1, 42-66542 (431). Lt. Chester D. Phillips was on a routine flight when he was reported missing at 1430 hrs., thirty miles south of Wau. Weather was the most probable cause of his loss. He also was from California.

September 13, 1943 P-38H5, 42-66749 (431). Lt. John C. Knox was a member of Blue Flight during a B-24 escort to Wewak when when it became separated from the rest of the squadron and ran into about ten Japanese fighters. Knox and Lt. Donno Bellows came to the rescue of the flight leader (Capt. Haning) when he was attacked by Tony fighters. Bellows and Knox were both shot down; Bellows returning to base on foot and Knox completely disappearing. The last words heard by Knox over the radio were something like, "Get those – off my tail."

September 13, 1943 P-38H5, 42-66734(432). Lt. Noel R. Lundy was the other 475th casualty on the B-24 escort to Wewak. He was leading the third Flight of the 432nd Squadron when he and his wingman, Lt. Webber, were attacked by three Tonys. The last Webber saw of Lundy's P-38 it was being raked along the belly by accurate fire from the Tonys. Lundy's body was found and buried near Bena Bena. He was officially declared KIA on January 18, 1944. He scored his last aerial victory on his last mission.

September 20, 1943 P-38H5, 42-66833 (432). Lt. Thomas J. Simms was apparently a member of the 432nd formation on an escort to Wewak. (The Squadron diary does not list him, but the after-mission report clearly identifies him.) Again, the squadron was jumped by Tony fighters and responded well by turning into the attack and claiming two Tonys and a Hamp. Simms was the only casualty when he was observed to bail out of his P-38. He was from Sacramento, Calif.

September 22, 1943 P-38H1, 42-66547 (432) Lt. Donald Garrison was the fifth 432nd pilot lost in a little over a month for the worst period in the squadron's combat history. He was in the flight led by Captain Harris during the big fight over the Finshhafen convoy. Lt. Vivian Cloud and Garrison were shot down during the fight which garnered the squadron eighteen confirmed claims. Cloud was rescued, but Garrison was never found.

September 24, 1943 P-38H, (433). Raymond P. Corrigan was one of the first 433rd Squadron pilot casualties. Apparently he was in contact with Japanese fighters in the Finschhafen area around noon when he was shot down. He was from sioux City Iowa.

September 24, 1943 P-38H1, 42-66580 (433). Lt. Kenneth D. Kirschner was the other 433rd Squadron pilot lost during this mission. He apparently escaped his downed P-38 about ten miles east of Finschhafen, but later died in captivity on Rabaul. He was from San Francisco.

October 17, 1943 P-38H1, 42-66561 (433). Lt. Virgil F. Hagan scored his first victory just two days before he was lost over Oro Bay. It is speculated that he was shot down by Japan's famous Zero ace, Nishizawa.

October 17, 1943 P-38H5, 42-66743 (431). Edward 1. Hedrick was also lost in the battle over the northeast coast of NG.

October 23, 1943 P-38H5, 42-66849 (431). Lt. Edward Czarnecki bailed out south of Cape Gazelle and made it to shore in his rubber dinghy. He was rescued in February 1944 by submarine, but ironically fell victim to a tuberculosis type of disease sometime later.

October 29, 1943 P-38H1, 42-66523 (432). Lt. Christopher Bartlett was flying the 432nd squadron commander's aircraft and was apparently shot down by Japanese fighters. His P-38 was found crashlanded with one engine shot out when the New Britain area fell into allied hands. Villagers nearby told of taking him in until Japanese troops were informed of his presence by a German missionary and he was taken to Rabaul where he apparently perished sometime early in 1944.

October 31, 1943 P-38H5, 42-66595 (432). Capt. Frederick A. Harris was testing Lt. William Ritter's P-38, which was not functioning properly, when he was seen to go down smoking two miles east of Cape Sudest in Oro Bay. Several search attempts were made to no avail.

November 2, 1943 P-38H1, 42-66665 (431). Lt. Kenneth M. Richardson was observed to be shot down in flames about three miles offshore from Rabaul. He apparently went down with his P-38.

November 2, 1943 P-38H5, 42-66821 (431). Lt. Lowell C. Lutton was last seen flying normally behind Lt. Arthur Wenige's White Flight. It is assumed that he ran out of gas on the way back from Rabaul. He was in the same Rankin Aeronautical Academy cadet class (42-A) as Dick Bong and his fifth air victory was confirmed on this fatal mission.

November 2, 1943 P-38H5, 42-66747 (432). Lt. Leo M. Mayo shot down a Tony and a portion of its wing flew back and tore off the P-38's right wing. Mayo managed to bail out and was seen to go ashore around Cape Mope, but he never returned from captivity.

November 2, 1943 P-38H5, 42-66857 (433). Lt. Donald Y. King went down over the Rabaul area and was listed as MIA. He had shot down his fourth Japanese aircraft on the previous Rabaul raid.

November 8, 1943 P-38H1, 42-66593 (431). Lt. Paul Smith MIA on a flight not believed to be combat related. He was an original pilot with the 475th from West Virginia.

November 9, 1943 P-38H1, 42-66596 (433). Lt. John C. Smith was an aggressive pilot who was lost on the same mission to the Alexishafen area as Danny Roberts and Dale Meyer. The wreckage of his plane was found in the same general area as Roberts' P-38. (See text)

November 9, 1943 P-38H5, 42-66834 (433). Capt. Daniel T. Roberts was lost in a collision with one of his squadron P-38s in the Alexishafen area. (See text.) Roberts' body was recovered and interred in the U.S. Memorial Cemetery on Luzon, P.1.

November 9, 1943 P-38H1. 42-66546 (433). Lt. Dale O. Meyer was also lost on the Alexishafen mission and was officially believed to have collided with Roberts. Subsequent evidence puts the issue into doubt. (See text.)

November 15, 1943 P-38H1, 42-66565 (432). Lt. Theodore Fostakowski took off on a five-muinute scramble at 12:23 in the afternoon for Nadzab. Somewhere enroute he developed some mechanical problem and returned to base. While trying to land on strip #5 he overshot the field and crashed about 400 yards off the west end of the strip as he attempted to go around again.

November 16, 1943 P-38H5, 42-66737 (431). Lt. Robert J. Smith was MIA on a flight that was probably not combat-related. He was from North Carolina.

December 28, 1943 P-38G-15, 43-2204 (431). Lt. Ormand E. Powell was MIA on a local flight, probably not the result of combat. He was from Richmond, Kentucky.

January 13, 1944 P-38 (probably) (432). Lt. John E. Fogarty was on a test hop in the late afternoon when he failed to pull out of a dive and crashed into the ground near the operations hut.

January 16, 1944 P-38H5, 42-66739 (433). Lt. Richard L. Hancock was MIA on a flight that probably was not combat-related. He was from San Francisco.

January 18, 1944 P-38H1, 42-66554 (432). Lt. William T. Ritter had a reputation as a well-liked eager beaver in the squadron. His zeal in combat was a bit too much on his last mission when he got too close to the Tony that he was shooting down and a piece of it flew back and ripped off his wing. Lt. John Michener was flying his wing and saw the P-38 go down in the Wewak area without a sign of a parachute. It was an incident eerily similar to the loss of Leo Mayo on November 2, 1943.

January 18, 1944 P-38H1, 42-66534 (431). Lt. John R. Weldon disappeared during combat over the Wewak area. One P-38 was seen spinning toward the jungle and may have been Weldon. Other reports mention seeing parachutes, but he never returned.

January 18, 1944 P-38H1, 42-66545 (431). Lt. Joseph A. Robertson was listed as killed in action when a drop tank from another formation hit his P-38 and forced him to take to his parachute. Even though he was seen to land in the jungle, subsequent reports must have confirmed his loss.

January 21, 1944 B-25(822 BS) 4264807 (431). Lt. McCleod Jones was acting as co-pilot on the B-25 on routine detached service when the bomber was listed as missing. Jones was never recovered.

January 22, 1944 P-38G, 42-12711 (431). Lt. Martin P. Hawthorne was killed in an aircraft accident, probably as a result of testing the P-38 rather than combat.

January 23, 1944 P-38H1, 42-66539 (433). Lt. Carl A. Danforth is listed as missing in action, but not believed due to combat.

January 23, 1944 P-38H5, 42-66852 (433). Lt. Donald D. Revenaugh also listed as MIA, but not believed due to enemy action.

February 14, 1944 P-38H1, 42-66577 (431). Lt. Wood D. Clodfelter had been having trouble with his own P-38 a day or two before he took off on his fatal flight. From whatever records have survived he apparently took off on an uneventful mission and crashed due to mechanical failure.

February 29, 1944 P-38, UNK (432). Lt. Harold Howard was in a flight of six P-38s heading for Cape Gloucester when poor weather forced two of the fighters to turn back and the other four to be completely wrecked. Howard was seen to bank sharply to avoid a moutain then catch a wing in the water and cartwheel into the sea.

March 31, 1944 P-38J10, 42-67801 (431). Lt. Robert P. Donald broke away from his element leader just as combat began in the Hollandia area. His right engine was smoking and he went into a spiral. Donald's element leader, Lt. Herman Zehring could not locate him again, but shot shot down a Japanese plane that may have been on an attack of Donald's P-38. According to other witnesses, the Japanese was shot off Donald's tail just before his P-38 spun into the ground.

April 3, 1944 P-38J10, 42-67584 (432). F/O Joe B. Barton crashed on the south end of Sentani Lake when a number of Tony fighters got on his tail and set his P-38 afire. He was seen to crash by the crew of an A-20 bomber.

April 16, 1944 P-38J10, 42-67605 (431). Lt. Jack F. Luddington listed as MIA on the Hollandia mission known as "Black Sunday." (See text)

April 16, 1944 P-38J15, 42-104355 (431) Lt. Milton A. MacDonald also listed as MIA on the "Black Sunday" mission (see text).

April 16, 1944 P-38J10, 42-67594 (432) Lt. Robert L. Hubner became disoriented on the flight and ceased to respond on the radio. Capt. Loisel advised the pilot to bail out, but the last ever heard of Hubner he was trying to get a radio fix.

April 16, 1944 P-38J15, 42-104390 (433). Lt. Louis L. Longman was last seen in a spiral tryinjg to break out of the weather between Hollandia and Saidor.

April 16, 1944 P-38J15, 42-104385 (433) Lt. Austin K. Neely, also seen turning out to sea in a spiral in the sudden bad weather.

April 16, 1944 P-38J15, 42-104381 (433) Lt. Lewis M. Yarbrough was the third 433rd pilot and sixth group loss of the day. Bob Tomberg also began to spiral away, but decided to bail out and walked back to camp.

June 11, 1944 P-38J10, 42-67593 (432) Lt. Troy L. Martin, started on an A-20 escort to Biak Island, but his wheels wouldn't retract. He circled Lake Sentani to drop his external tanks and his engines cut out, causing his P-38 to crash into a swamp. Martin was from Twenty-nine Palms, California.

June 16, 1944 P-38J15, 42104364 (433). Lt. Howard V. Stiles had been lucky several times before, including the "Black Sunday" mission of April 16, 1944. He was part of an eleven plane escort of B-25s over Jefman and Samate when a group of Oscars and Tojos was engaged. He was last seen chasing an Oscar on the deck in Jefman Harbor. He was from Newton, Massachussetts.

June 30, 1944 P-38J10, 42-67793 (431). Lt. Robert L. Crosswait was probably on one of the 475th's first divebombing missions to the Noemfoor Island area. There is no mention in records whether he had mechanical trouble or was hit by enemy fire, but he is listed as MIA. He was from Kinder, Louisiana.

July 28, 1944 P-38J10, 42-67787 (432). Lt. William A. Elliot took off from Noemfoor at 8:30 in the morning for Biak after spending the night. He had landed there after the Halmaheras strike of July 27 because of mechanical difficulty. At 11:00 he called in saying that he had only five minutes fuel left and was going to ditch. Searches failed to find any sign of him. He was from San Mateo, California.

August 4, 1944 P-38J15, 42-104161 (431). Capt. William S. O'Brien had been an early member of the 431st and died in a freak combat incident. He was part of a sixteen plane flight that included Charles Lindbergh in a strike at Liang airstrip. Leading Red Flight, O'Brien engaged a Tony that looped back and crashed head on into him. Both planes fell into a mass of flames and crashed in the water. The Tony was awarded to O'Brien as his fourth kill.

September 19, 1944 P-38J15, 42-104037 (431). Lt. Nathaniel V. Landen is listed as KIA, but it is not believed that he was on a combat mission at the time of his loss. It is probable that he crashed during a routine cross country or training flight. He was from Live Oak, Florida.

September 24, 1944 P-39Q, 42-20025 (432). Lt. Walter W. Weisfus took off at 1445 on a local flight in a P-39 borrowed from the 66th Service Squadron. He had enough fuel for a ninety minute flight and couldn't be contacted after 1615. Searches were made for days afterward, but he was never found. He was from Chicago.

October 1, 1944 P-38J20, 44-23551 ((433). Lt. Charles H. Joseph began a divebombing mission to Fak Fak when his P-38 crashed on takeoff and he was killed. His hometown was Greensburg, Pennsylvania.

October 2, 1944 P-38L1, 44-23958 (432). Capt. Billy M. Gresham had been one of the hottest pilots of the original 432nd cadre. He would undoubtedly have gone home before long,but was lost in another freak accident. Apparently he was trying out one of the new P-38L aircraft with aileron boost and compressibility flaps when he became overdue. Controller could not reach him on any channel. His P-38 was found crashed northwest of Borokoe drome on Biak. His body was found nearby with the parachute only partially deployed.

October 13, 1944 P-38J15, 42-104343 (431). Lieutenant Donald W. Patterson reported MIA on this date, near Noemfoor, due to a training accident.

November 3, 1944 P-38L1, 44-24172 (431). Lt. Arnold R. Nielson collided with an aircraft from another group at 11:30 am in a landing approach over the congested Tacloban strip.

November 10, 1944 P-38L1, 44-23935 (432). Lt. Grady M. Laseter collided with Lt. Perry Dahl over Ormoc Bay. Dahl survived,but Laseter was never returned.

November 24, 1944 P 38L1, 44-24888 (431). Lt. Erling J. Varland was flying in a four-plane flight that was patroling shipping in the Leyte Culf. He was in no apparent trouble when he was last seen and simply vanished from the flight.

December 7, 1944 P-38L1, 44-23945 (432). Lt. Col. Meryl M. Smith was involved in a number of actions during the most hectic day of aerial combat over Leyte. At about 2:30 in the afternoon over Ponson Island a number of Mitsubishi Jack fighters attacked his flight and Smith was seen to go down after being hit by an enemy fighter.

December 6, 1944 P-38L1, 44-24858 (433). Lt. Morton B. Ryerson reported crashed at 0925, three miles southwest of Dulag, probably on morning mission.

December 25, 1944 P-38L1, 44-24846 (431). Lt. Robert A. Koeck last seen in shallow dive over Clark Field.

December 25, 1944 P-38L1, 44-24889 (431). Lt. Enrique Provencio was last seen in his flight over Clark at 1045.

December 31, 1944 P-38J15, 43-28829 (433). Lt. Clifford L. Ettien probably developed some sort of mechanical problem during an escort of C-47s to Mindoro. He was observed to bail out too low for his survival about two to three southeast of Masbate at 4:10 in the afternoon.

January 2, 1945 P-38L1, 44-24843 (432). Capt. Henry L. Condon II was the extremely popular commander of the squadron and was leading an escort mission to Porac and Floridablanca when he went down to strafe a Japanese troop train north of Manila. The engine exploded and perhaps damaged Condon's P-38. He tried to bail out, but his aircraft crashed and blew up with him still inside.

January 7, 1945 P-38L1, 44-24845 (431). Major Thomas B. McGuire, Jr. was lost during a fighter sweep over Negros Island. (See text)

January 7, 1945 P-38J15, 43-28836 (431). Major Jack B. Rittmayer flying with the 431st as a member of McGuire's flight was shot down by Fukuda's Ki-84 Frank. (See text)

January 15, 1945 P-38L5, 44-25336 (432). Capt. Paul W. Lucas was on patrol over a field twenty to thirty yards south of a river that was about one mile north of LaCarlotta strip on Negros when he was hit by light AA and machine gun fire in the right engine. His P-38 went down with both engines apparently on fire. P-38 bounded about fifty feet on impact and the belly tank went up in flames in addition to tail breaking off. Philippinos were seen running up to the slightly burning plane and the pilot was slumped in the cockpit.

January 29, 1945 P-38J15, 43 28554 (431). Lt. Robert Patterson was reported MIA, probably during an operational mission.

January 29, 1945 P-38J15, 43-28551 (432). F/O Charles C. Nacke was MIA about fifteen miles northwest of Dulag strip. A flash on the ground was observed and there was speculation that he bailed out unsuccessfully.

February 15, 1945 P-38L5, 44-25428 (431). Lt. Arthur J. Schmitt was MIA on a mission over Panay Island.

March 28, 1945 P-38L5, 44-25459 (433). F/O Charles R. DeWeese crashed 40 miles west of Cabra Island. Apparently, bail out attempt was unsuccessiul.

March 28, 1945 P-38L5, 44-25498 (433). Lt. Wesley J. Hulett ditched three to four miles southwest of Tre Island off the coast of Indo-China when both engines were shot out by Japanese fighter. His comrade, F/O Bert Simmons, flew a ten hour night search mission without success.

April 5, 1945 P-38L1, 44-24844 (433). Lt. Laverne P. Busch was assigned to the group in March 1945. He was on a mission to Clark Field in Lt. William Richmond's flight when contact was lost enroute to target, April 2, 1945. Apparently, Busch's napalm bomb was hungup and he attempted to bail out. He was recovered with multiple fractures and shock, succumbing on April 5.

April 18, 1945 P-38J20, 44-23372 (432). Lt. Reed L. Pietscher of Princeton, Iowa, was on the fourth divebombing mission of the day, crashed into the side of a mountain southwest of Minuli, Luzon, on highway Route #5. Some of the other pilots on the mission thought that he may have been hit by ground fire as holes were observed on the underside of his P-38.

May 30, 1945 P-38L5, 44-25628 (431). Lt. Millard R. Sherman was on a low-level mission to Taiwan when his P-38 struck a tower, building or tree about three miles southwest of the town of Sank Yaku.

June 18, 1945 P-38L5, 44-26303 (431). Lt. Alvin G. Roth was last seen at very low altitude in his burning P-38 during mission to Taiwan. Apparently he was hit by small arms fire in the right wing tank and was only partly out of the P-38 when it crashed into the water.

June 18, 1945 P-38L5, 44-25414 (431). Lt. Edward Carley was on the same mission to Taiwan. He was also hit by small arms fire and was seen to attempt a bail out around the mouth of the Subun-Kei River, but apparently he was too low. He was never found by the rescue planes that looked for him.

June 21, 1945 P-38L5, 44-25312 (431). Lt. Herbert S. Finney was apparently part of an operational flight when he crashed on takeoff and was KIA.

July 8, 1945 P-38L5, 44-26473 (433) Lt. Charles G. Zarling was another young pilot who had been with the group only since March. He was on a local flight and was seen to dive at a 60 degree angle into the Lingayen Gulf, one-half mile north of the air-strip. He was from Lawton, Oklahoma.

July 11, 1945 P-38L5, 44-25968 (HQ). Capt. George W. Smith was flying in a four-ship training flight with the 431st Squadron. He was the number three man in the flight and was seen to half roll and crash into the water. His wingman tried to contact him, but later speculation suggested that he was suffering from vertigo after the flight emerged from a cloud.

Also:
February 22, 1945: A runaway P-47 plowed through the 433rd parking area killing 3 crewchiefs:

SSgt Charles Huff, Pittsburgh, PA
Sgt Edward J. Hamilton, Allendale, NJ
Cpl. William E. Maddock, Philadelphia, PA

475th Fighter Group Aircraft

Serial Number	Pilot	Date and Comment	Serial Number	Pilot	Date and Comment
					Zero, Hamp 10/23/43
P-38F-2					Ret. early 11/02/43
42-12652		probably original training equipment		Cohn	Ret. early 10/29/43
42-12653	Ekdahl	Return early 1/18/44		R.J. Smith	Ret. early 10/24/43
42-12711	Hawthorne	MIA 1/22/44		R. Herman	Zero 11/07/43
			42-66672	Benevant	Ret. early 1/18/44
P-38G-15					
43-2204	Powell	MIA 12/28/43	**P-38H-5**		
			42-66737	Lewis	Oscar 8/16/43
P-38H-1				R. Smith	MIA 11/16/43
42-66502*	Czarnecki	2 Zero 8/16/43; Nate 8/21/43	42-66742	Jett	2 Zero 8/18/43
		AC left U.S. June 1943 and probably was		Brown	Zero 10/24/43
		returned Oct 1945.		Nichols	Ret. early 11/01/43
42-66511	Wenige	2 Zero 8/16/43; 2 Zero 11/02/43; Tony		Morriss	Escort Elliott 11/02/43
		11/07/43	42-66743	Hedrick	MIA 10/17/43
42-66522	Sieber	Return early 9/13/43; missing 10/15/43 (ret)	42-66744	Allen	Oscar,Kate 8/16/43; 2 Zero, Oscar 8/21/43
	O'Brien	Ret. early 10/12/43 (gas siphon)		Monk	Zero 10/15/43
42-66524*	Monk	Ret. early 10/23/43 (severe illness)		Cline	Damaged in combat 10/29/43
42-66531	Mankin	Ret. early 10/12/43 (eng. overheat)	42-66746	Brown/Monk	#118 "Petty Pretty"
	Riegle	Badly damaged 10/15/43		Dunlap	Hung drop tank 9/13/43
42-66534*	Zehring	Ret. early 12/22/43		Lutton	Zero 10/23/43
	Weldon	MIA 1/18/44		Champlin	2 Zero 11/02/43
42-66537*	Lewis	Oscar 8/18/43; Zero, Prob. 11/02/43		Monk	Ret. early (propeller out) 1/18/44
42-66540	Nichols	Oscar 8/21/43	42-66749	Knox	MIA 9/13/43
42-66542	Phillips	MIA 9/8/43	42-66751	Haning	Zero (Prob) 8/18/43
42-66545	Robertson	MIA 1/18/44	42-66764	Bellows	#130 "Piss Pot Pete"
42-66548	Lutton	Oscar 8/16/43	42-66817	McGuire	#131 "Pudgy II"
	Samms	Ret. early 11/02/43	42-66821	Houseworth	2 Zero 10/12/43.(prob. Zero also)
	Smith	Ret. early 11/07/43		Lent	Hydraulic system out 9/28/43
	MacDonald	Reportedly flown by him on mission of		Lutton	MIA 11/02/43
		1/18/44. One Hamp with 432nd Squadron.	42-66825	Jett/Herman	#120 "Thoughts of Midnite"
42-66550	Elliott	Zero 10/17/43	42-66826	Morriss	Gunsight out 10/12/43
	Mankin	Oscar, Tony 11/07/43			Zero 10/15/43
					Escort Monk 10/23/43
			42-66827	Kirby	2 Zero 11/02/43
P-38H-1			42-66828	Morriss	Return with Hydraulic system out 9/28/43
42-66554	R. Herman	Left intercooler shot out by AA, 10/23/43		Hunt	Return early from Kiriwina 11/02/43
42-66556		Reported to have crashed in the vicinity of		Veit	Escort Ekdahl 2/13/44
		Rogers A/D 9/5/43	42-66836	McGuire	Shot down 10/17/43; #110, Nichols, P-38
42-66558	Donald	Crashlanded at Base 12/22/43	42-66837	Samms	Crashland Ramu Valley 9/13/43
42 66572	Schmidt	MIA 8/18/43	42-66840	Gronemeyer	#121 "Little Grace"
42-66577	Clodfelter	MIA 2/14/44	42-66843	Giertsen	MIA(Return) 11/02/43; originally assigned
42-66589	Jett	Damaged by 20 mm cannon fire 1/18/44			to Dean of 432nd Squadron
42-66592	Brown	3 Zero 8/16/43	42-66846	MacDonald	2 Oscars 11/09/43
	Czarnecki	Ret. early, loss of power 10/17/43	42-66849	Czarnecki	MIA 10/23/43
	R.J. Smith	Escort Houseworth 10/23/43	42-66853	Hawthorne	Spare 1/10/44
	Monk	Zero 11/02/43	42-66902	Wenige	Hamp(prob) 10/23/43
42-66593	Lutton	2 Oscar 8/18/43	42-66906	Ballard	Ret. early 1/18/44
	Smith	MIA 11/08/43	42-66908	Champlin	10/12/43
42-66594	Smith	Damaged in action 10/29/43		Czarnecki	2 Zero 10/15/43
	Elliott	Val 10/15/43		Pare	Crashlanded 10/17/43
42-66665	Richardson	MIA 11/02/43	42-66912	Lent	Ret. early (fuel trouble) 12/22/43
42-66666	Elliott	Ret. early 10/12/43 with hung belly tank;		Ekdahl	Ret. early 2/13/44

Served in more than one 475th Squadron

Serial Number	Pilot	Date and Comment

P-38J-10

Serial Number	Pilot	Date and Comment
42-67143	Champlin	Engine cut out on XC to Wadke 6/16/44
42-67580	Provencio	Ret. early 6/6/44
42-67586	Sidnam	Both engines cut out, local test 3/6/44
42-67588	Sexton	Ret. early 6/6/44
42-67597	Champlin	#113 "Buffalo Blitz", Ret. early 2/15/44
42-67605	Hawthorne	MIA 1/22/44
42-67801	Donald	KIA 3/31/44
42-67802	Maxwell	Nose tire blown, Trn to 482 Serron 4/12/44
42-67807	Ballard	Bomb fin lodged in horiz. stab near Liang 9/9/44, to 66 Serron 9/13/44

P-38J-15

Serial Number	Pilot	Date and Comment
42-103855	Kidd	Ret. early 6/16/44
42-103965	Barnes	Ret. early 9/19/44
42-103982	Loe	Crashlanded on trng. mission 4/27/44
42-103991	Roark	Crashland Nadzab strip #3 4/22/44
42-103992	Knecht	Ran out of fuel making emergency landing at Hollandia 6/16/44
42-103996		
42-104019	Reeves	Tire blown on landing 5/20/44, to 61st Serron
42-104020	Cohn	Ret. early 6/16/44
	Tilley	Landing Gear wouldn't retract 8/26/44
42-104021	Gronemeyer	#121 "Little Grace II"
	Hart	Abort 6/16/44
42-104022	Monk	Left gear collapsed on Finschhafen strip 3/22/44
42-104025	Veit	Belly tank hit fuselage over Jefman 6/16/44; to 46th Service Gp. 6/17/44
42-104028	Zehring	Ret. early (Prop out) 6/4/44
42-104030	Madison	Hit mast of Fox Tare while strafing 6/6/44; to 46th Ser. Gp.
42-104032	Lent	#134 "T-Rigor Mortis II"
	Benevent	Ret. early 6/16/44
42-104034	Moering	Ret. early 9/19/44
	Pearson	Bomb dropped on landing 9/16/44
42-104037	Landen	MIA 9/19/44
42-104038	Cline	#135 "Pappy's Bir-rdie" Ret. early 6/16/44

P-38J-15

Serial Number	Pilot	Date and Comment
42-104126	Cortner	#117 Ret. early 3/31/44
42-104161	O'Brien	MIA 8/04/44
42-104286	Moering	Ret. early 12/20/44
42-104343	Provencio	Ret. early 9/19/44
	Patterson	MIA 10/13/44
42-104355	MacDonald	MIA 4/16/44
42-104387	(on ground)	F5 crash into revetment area 7/29/44
42-104911	Sidnam	Crashlanded 6/16/44; to 61st Serron
43-28525	Thropp	Damaged in boom and engine 1/7/45
43-28554	Patterson	MIA 1/29/45
43-28836	Rittmayer	MIA 1/7/45
43-29042	(on ground)	F5 crash into revetment area 7/29/44

P-38J-20

Serial Number	Pilot	Date and Comment
44-23291	Champlin	Belly tank hit gondola on Liang mission 9/9/44; to 66th Serron 9/11/44
44-23294	Morriss	Ret. early 9/19/44
44-23296	Fulkerson	Ret. early 12/20/44
44-23312	Madison	Ret. early 5/5/45
44-23362	Weary	Crashlanded Lingayen Gulf 5/28/45
44-23379	Rohrer	Crashlanded 5/1/45

P-38L-1

Serial Number	Pilot	Date and Comment
44-24172	Nielson	Collided with another aircraft 11/03/44
44-24187	Pietz	Ret. early 5/28/45 (right wheel would not retract.)
44-24845	Gronemeyer	#112 "Eileen-Anne" or "Kim" Ret. early (low fuel) 12/6/44
	McGuire	KIA 1/7/45
44-24846	Koeck	MIA 12/25/44
44-24847	Parshall	Ret. early 12/6/44
44-24848	Pearson	Ret. early 1/1/45
44-24850	(on ground)	Damaged on Tacloban Strip by bomb 11/15/44; to 96th Serron
44-24869	Tilley	AA damage in wing, Nov. 44; to 479th Serro
44-24871	Pietz	Damage from ret. fire of Japanese bomber 11/24/44; to 479th Serron
44-24874	Fulkerson	2 Jack 12/25/44, shot down same day and pilot returned.
44-24875	(on ground)	Damaged on Tacloban Strip by bomb 11/13/44; to 392nd Serron
44-24853	Herman	Ret. early (fuel cap loss) 12/20/44
44-24888	Varland	KIA 11/28/44
44-24889	Provencio	MIA 12/25/44

P-38L-5

Serial Number	Pilot	Date and Comment
44-25130	Turner	Ret. early 5/4/45
44-25224	Turner	Ret. early 5/21/45
	Holdren	Ret. early 7/10/45
44-25303	Max	Ret. early 6/19/45
44-25307	O'Neill	Ret. early 5/26/45
	Jones	Ret. early 6/17/45
44-25312	Finney	KIA 6/21/45
44-25332	Hanway	Ret. early 2/25/45
	Smith	Ret. early 5/5/45
44-25339	Sherman	Damaged over Clark 4/18/45
44-25428	Schmitt	MIA 2/15/45
44-25439	DuMontier	#135 "MADU V"
44-25446	Weaver	Ret. early 5/14/45
44-25447	Olson	Ret. early 4/6/45; 5/12/45
44-25482	Werth	#139 "Irish Angel"
	Parshall	Ret early 5/5/45
	Fenton	Ret early (low fuel pressure) 5/28/45
44-25628	Sherman	MIA 5/30/45
44-25635	Werner	Escort Hanway 5/19/45
	Holdren	Ret. early 4/6/45
44-25639	Pietz	#126 "Vickie"
44-25648	DuMontier	#135 "MADU IV"
44-25650	Box	Bail out over Clark, AA damage 4/18/45
44-25653	Hart	Ret. early (Hydraulics out) 4/24/45
44-25656	Gray	#112 "KIM IV"
	Harris	Damaged stabilizer from 20mm fire 5/28/45
	Bratton	Ret. early 6/7/45
	Fenton	Ret. early 6/13/45
44-25868	Weaver	Damaged over Clark 4/18/45
44-25878	Oxford	#122 "Doots II"
44-26187	Pietz	Ret. early 5/28/45 #115
	Reeves	Ret. early 6/19/45
44-26303	Roth	KIA 6/18/45
44-26401	Neal	#116 "Miss Step ins"
44-26404	Fenton	Ret. early 6/19/45
44-26393	Giertsen	Crashed on TO for test hop 2/1/45

Serial Number	Pilot	Date and Comment	Serial Number	Pilot	Date and Comment

432nd Fighter Squadron

P-38G-5

42-12644	Hedrick	Ret. early 12/16/43	42-66855	Tomkins	Ret.early 12/13, 12/16/43
			42-66879	Hedrick	20 mm damage 12/22/43
P-38G-15			42-66905	Fostakowski	Oscar(prob) 11/02/43
42-2280	Beatty	Ret. early 12/16/43	42-66913	Forster	Ret. early 12/16/43

P-38H-1

42-66502*	Dean	Betty, Zero 9/22/43; Val 10/15/43; 2 Zero 10/24/43; 2 Zero 12/22/43	**P-38J-5**		
			42-67140	Dickey	Ret. early 4/03/44
42-66504	Dahl	Zero 11/09/43	42-67141	Schuh	AA damage right boom & vert. stab. 9/03/44; to 66th Serron
42-66523	Wilson	#140; First operational P-38 assigned to the 432nd Squadron.	42-67144	Col. MacDonald	Ret. early (Dropped belly tank) 1/23/44
	Bartlett	MIA 10/29/43	42-67584	Barton	MIA 4/03/44
42-66513	Roberts	2 Hamp 8/21/43; Oscar 9/07/43	42-67593	Lawhead	Ret. early 4/03/44
	Michener	Ret. early 12/13/43		Martin	KIA 6/11/44
42-66534*	Condon	Zero 9/22/43; Betty 9/30/43	42-67594	Hubner	MIA 4/16/44
42-66537*	Loisel	2 Tony 8/21/43; Zero 9/22/43	42-67589	Lamb	Broken nose strut, end of Hollandia strip hit by incoming P-40 5/29/44
	Rundell	2 Zero 11/09/43			
42-66541	Summer	Zero 10/24/43	42-67599	Blakely	Crashland 5 mi. off Cape Waios 7/27/44
	Ince	Betty, Zero 9/22/43; Zero 11/09/43	42-67601	Roberts	Injured in Hollandia crashlanding 6/44
42-66547	Garrison	MIA 10/17/43	42-67787	Bradley	Damaged in aerial combat 6/16/44
42-66553	Harris	2 Tony (4 Dam.) 8/21/43; Zero (Zero Prob.) 10/24/43			
	Cloud	Ret. early 11/16/43	**P-38J-15**		
	Barton	Ret. early 2/15/44	42-104026	Loisel	Slight damage 7/27/44
42-66554	Fogarty	Betty 10/12/43	42-104029	Wilson	Damaged 5/19/44; to 61st Serron
	Ritter	MIA 1/18/44	42-104035	Summer	Broken nosewheel on landing at Wakde 6/07/44; to 46th Serron
42-66555	Lucas	Zero 8/21/43			
	Allen	Ret. early (Late takeoff) 1/23/44	42-104036	(on ground)	Hit by schrapnel from enemy bomb on Wakde 6/09/44; to 100 Serron
42-66565	Gresham	2 Tony 9/20/43; 100 hr svc check 9/25/43			
	Fostakowski	MIA 11/15/43	42-104350	Likins	L/C folded on landing at strip #3 Nadzab 4/8/44
42-66575	Summer	Zero 8/21/43			
	Cloud	Betty, Oscar 9/22/43; Val 10/15/43	42-104376	Gresham	Crashlanded Hollandia strip on local test 6/21/44
42-66568	Ince	Oscar 11/09/43; Oscar 11/16/43			
	Gresham	Zero 10/24/43	42-104388	Miller	L/G folded on landing at strip #3 Nadzab 4/7/44
42-66580	Kirshner	MIA 9/24/43			
42-66595	Condon	Betty 9/30/43	43-28493	Koles	Left engine blew up near Dulag 11/27/44; Trnsf. to 479th Serron
	Harris	MIA 10/31/43			
42-66632	Hansen	Ret. early 12/13/43	43-28551	Nacke	MIA 1/29/45
	Michener	Ret. early 1/23/44			
42-66633	Gholson	T/E 8/21/43	**P-38L-1**		
	Hadley	Ret. early (oil pressure) 1/23/44	44-23372	Pietscher	MIA 4/18/45
42-66682	Bartlett	Flown in from Eagle Farms 9/14/43	44-23921	Condon	Struck by own belly tank on B-24 escort 12/23/44
	Loisel	2 Zero 10/15/43			
				(on ground)	Damaged in collision with 433rd P-38 1/27/45
P-38H-5					
42-66734	Lundy	Zero 9/7/43; Tony 9/13/43; MIA 9/13/43	44-23930	Willis	Shot up by enemy fighters 11/10/44 to 96th Serron
42-66747	Mayo	Tony, MIA 11/02/43			
42 66748	Ryholm	MIA 9/4/43	44-23935	Laseter	KIA 11/10/44
42-66750	Gresham	T/E 8/21/43; Ret. on single-engine 9/20/43	44-23938	Willis	Damage nosegear on muddy San Pablo #2 11/6/44
	Peregoy	Zero 11/02/43			
42-66832	Harris	2 Val 10/15/43	44-23942	(on ground)	Destroyed by schrapnel 11/16/44; to 10 Serron
42-66833	Simms	MIA 9/20/43			
42-66842	Farris	Damaged 10/18/44	44-23945	M.M. Smith	MIA 12/7/44
42-66818	Summer	Oscar 10/15/43	44-23948	(on ground)	bombed on ground, complete loss 11/16/44
	Dahl	Ret. early (Left engine out) 11/16/43, damaged 11/19/43	44-23950	(on ground)	bombed on ground 11/16/44; to 10 Serron
			44-23957	Dahl	Lost 11/10/44; pilot returned
	Howard	Ret. early 2/15/43	44-23958	Gresham	MIA test flight 10/02/44
42-66844	Cloud	Val 10/15/43	44-23964	–	Records indicate in service with 432nd 11/44 to 12/45
	Forster	Zero (prob) 11/16/43			
42-66846	Gresham	Kate 10/15/43	44-23975	Blakely	Damaged 11/11/44
	Rundell	Betty 9/22/43	44-24114	Summer	Wing scorched 11/12/44
			44-24115	Schuh	Crashlanded wheels up 12/12/44, to 479th Serron

Served in more than one 475th Squadron

Serial Number	Pilot	Date and Comment
44-24121	Hannan	10/09/44 Wing struck tree on test flight — no injury
44-24152	(on ground)	Damaged by bombs on San Pablo strip #2 11/16/44; to 10th Serron
44-24166	Morris	Nosewheel collapse on landing at Dulag 12/23/44
44-24843	Condon	MIA 1/2/45
44-24848	(on ground)	Damaged by bombs on San Pablo strip #2 11/16/44
44-24850	(on ground)	

P-38L-5

Serial Number	Pilot	Date and Comment
44-25314	Summer	Crashland 2/19/45, faulty landing gear
44-25336	Lucas	KIA 1/15/45

Serial Number	Pilot	Date and Comment
44-25412		Crewchief SSgt Calvin. Burst into flames during carburetor blending 4/19/45; Calvin injured and P-38 to 10th Serron
44-25459	DeWeese	MIA 3/28/45
44-25479	Blakely	Damaged 3/28/45
44-25492	Goodwin	Bail out north central Lingayen coast 6/6/45
44-25564		Reported as taken on charge Jan/Feb 1945
44-25600	Frazee	Electrical loss on trng. flt. Complete loss 7/20/45
44-25643*	Loisel	Damaged in taxi collison 1/27/45; to 10th SS.
44-25640	Williams	Schrapnel damage 3/25/45
	Dowler	Damaged in both props by e/a mg 3/29/45
44-25971	Owen	"Betty Jo"

433rd Fighter Squadron

P-38H-1

Serial Number	Pilot	Date and Comment
42-665		
42-66515	–	Maintenance check (100 hr?) 12/08/43
42-66521	McKeon	Oscar 9/04/43
	Smith	2 Zero 10/17/43
	Lewis	Ret. early 11/05/43
42-66524**	Reinhardt	To Rabaul 10/24/43
	Jeakle	Damaged in noseover 12/18/43
42-66533	Smith	Zero 10/29/43
	Tomberg	To Rabaul 11/02/43
	Lewis	Hamp (prob. damage) Tony (Prob.) 11/09/43
42-66538	Smith	Hamp, Dinah (prob.) 9/02/43
42-66539	Meyer	To Rabaul 10/23/43
42-66546	Grice	Zero 10/17/43, Ret. early 10/23/43
	Cochran	Ret. early 11/05/43
42-66551	Brenizer	To Rabaul 10/23/43
		Damaged 11/16/43
		Damaged in midair 12/18/43
42-66560		Damaged 10/15/43
42-66561	Hagan	MIA 10/17/43
42-66566	Longman	To Rabaul 10/24/43
42-66574	Neely	To Rabaul 10/23/43
	Fisk	Zero 10/24/43
	Roberts	Zero 10/29/43
42-66579	Northrup	To Rabaul 10/24/43
	Howard	Zero 10/17/43
42-66580	Kirschner	MIA 9/24/43
42-66587	Ehlinger	Ret. early 11/07/43
	Cleage	Cleage Oscar 10/12/43
	Peters	Peters To Rabaul 10/23/43
	King	Zero, Oscar 10/24/43
42-66589	Grady	Damaged 10/15/43
	Danforth	To Rabaul 11/02/43
	Howard	Ret. early 11/16/43
42-66596	Northrup	Ret. early 11/05/43
	Smith	MIA 11/09/43
42-66598	Hagan	Oscar 10/15/43

Serial Number	Pilot	Date and Comment
42-66634	Longman	Ret. early 11/05/43
42-66636	Knecht	S/D 2 mi. s. of Sigul River, Arawe, N.B. 12/16/43; Returned uninjured.

P-38H-5

Serial Number	Pilot	Date and Comment
42-66739	Meyer	To Rabaul 10/23/43
42-66752	Roberts	2 Zero 10/23/43
	Palmer	Return early 11/07/43
	Badgett	MIA 2/8/43
42-66753	Palmer	To Rabaul 10/23/43
	Wire	To Rabaul 10/24/43
	Neely	To Rabaul 11/02/43
	Cochran	Hamp Prob. 11/09/43
42-66834	Roberts	MIA 11/09/43
42-66839	Cochran	Zero 10/24/43
	Grady	Ret. early 11/05/43
42-66843	Cochran	Zero 10/29/43
42-66852	Fisk	#189
	Roberts	2 Zero 10/17/43
	Malloy	Val, Val Prob. 10/15/43
	Revenaugh	MIA 1/23/44
42-66853	Fisk	2 Zero 10/17/43
42-66854	King	#177
	Lewis	Ret. early 10/29/43
	McKeon	Hamp 9/24/43; Val 10/15/43
42-66856	Danforth	To Rabaul 10/23/43
42-66857	Revenaugh	Damaged 10/15/43
	Smith	Oscar 10/15/43
	King	MIA 11/02/43

P-38J-5

Serial Number	Pilot	Date and Comment
42-67282	–	#172

P-38J-15

Serial Number	Pilot	Date and Comment
42-104003	–	C/c DeBeck #174
42-104024	Col. MacDonald	C/c Dietz #100 maintained Feb. Aug. 1944
42-104015	Tomberg	#177
42-104035	Lewis	#170 C/c Rath

* Col. MacDonald's PUTT-PUTT MARU subsequently repaired and assigned to 8th FG.

** Served in more than one 475th Squadron

Serial Number	Pilot	Date and Comment
42-104302	Jeakle	C/c Weaver #173
42-104310	Northrup	C/c McConnell #175
42-104286		C/c Van Horn #172
42-104298		C/c Gross #192
42-104337	Knecht	Crashlanded 4/17/44
42-104342		C/c Guinn #177
42-104343		C/c Landwehr #179
42-104345	Ryerson	C/c Hamilton #180
42-104351	Wire	
42-104352	Price	Crashlanded
42-104354		C/c Kline #183
42-104357		C/c Noseck #184
42-104358	Cochran	Damaged beyond repair in accident 7/28/44
42-104359	Hasty	S/D 6/05/44; POW
42-104364	Stiles	MIA 6/16/44
42-104360		C/c Baun #190
42-104370	Heath	#197
42-104374		C/c Stiglica #191
42-104381	Yarbrough	MIA 6/16/44
42-104380		Left U.S. Feb. 22, 1944; in 433rd end of war
42-104385	Neely.	MIA 6/16/44
42-104390	Longman	MIA 6/16/44
		C/c Anderson #185
42-104494	Brenizer	#182 "Chase's Ace"
42-104497		C/c Senkler #171
42-104501		C/c Coleman #181
42-104502		C/c De France #193
42-104506		C/c Snyder #195
42-104508	Anderson	C/c Ruiz #194
42-104840		C/c Sanders #198
42-104995		C/c Massengale #196
43-23863	(on ground)	Destroyed in P-47 takeoff crash 12/04/44
43-28505	Sperling	Crashland Mangarin Bay
43-28561	Hulett	Hit by AA (hydr., right wing) 9/10/44; to 66th Serron

P-38L-1

Serial Number	Pilot	Date and Comment
44-23300		Hit in right wing with nil effect, 2/14/45
44-24841	Nelson	Damaged by AA over Cebu Bay 12/5/44
44-24842	Purdy	Crashlanded, total loss 12/11/44
44-24844	Busch	KIA 4/05/45
44-24851	(on ground)	Destroyed in P-47 takeoff crash 12/04/44
44-24855	Simpson	Damaged Formosa 6/02/45
44-24858	Ryerson	Damaged 11/24/44; destroyed in crashlanding west of Dulag strip.
44-24859	(on ground)	Destroyed in P-47 takeoff crash 12/04/44
44-24863	(on ground)	P-47 crash on takeoff Dulag strip 12/04/44
44-24864	(on ground)	P-47 crash on takeoff Dulag strip 12/04/44
44-24865	(on ground)	P-47 crash on takeoff; Nose damage; to 479th SS
44-24872	Wire	Damaged on Leyte Gulf patrol 1/06/45; to 10th Serron
44-24890	Ross	Scramble 11/24/44; S/D by e/a and crashlanded on reef; rescued; paper transfer of P-38 to 46th Serron

P-38L-5

Serial Number	Pilot	Date and Comment
44-25131		Ret. early (damaged aileron boost system) 2/14/45
44-25148	Ralph	Nosewheel door torn off 7/23/45; to 64th Service Group
44-25498	Hulett	MIA 3/29/45
44-25657	T.F. Smith	Crashland on road west of Hill strip 3/01/45; complete loss
44-25880	Morrell	Belly tank hit gondola over west coast of Formosa 4/06/45; to 66th Serron
44-25930	Purdy	#174 "Lizzie V" crashland 1/09/45
44-26278	Williford	Belly tank hit tail over southwest Canton, China 7/13/45; to 96th Serron

Loisel's #100 at the end of the war (Ethell)

#101 at about the time of the Philippine campaign. (Ethell)

#102. (Krane)

(Krane collection)

(Krane collection)

(Prell)

Lt. Martin's #114. (Krane)

LITTLE LORRENE, right side of #115, MISS FLUFF'N LACE.

(Tabatt)

(J. Walker)

(Krane collection)

(Krane collection)

(Krane collection)

(Krane collection)

(Krane collection)

(Krane collection)

(Krane collection)

(Krane collection)

(Krane collection)

(Krane collection)

(Krane collection)

#127. (Prell)

(Krane collection)

(Krane collection)

(Prell)

(Brown via Anderson)

#135 was variously used by Major Bob Cline, Major John Vogel and Lt. Bob Weary in 1945. (Krane collection)

MRS BILL was originally 102, but became 138 in the 431st Sq. (Krane collection)

Henry Condon on the wing of John Loisel's #140.(Krane coll.)

Summer's #140. (Krane collection)

#141 about the end of 1943. (Krane collection)

#141 in the latter part of 1944. Note 80th Sq. P-38 in foreground. (Krane collection)

P-38 of Lt. Leo Blakely (Krane collection)

Art Peregoy's #159, serial 42-66850, SLIGHTLY DANGEROUS. (Krane collection)

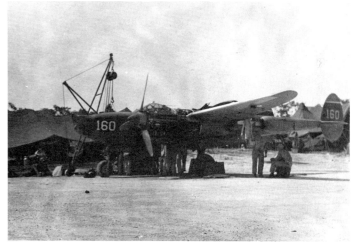

Lt. Ferdinand Hanson's #160, P-38J-5, Serial 42-67290. (Krane collection)

#163 in mid-1944. (Krane collection)

Front quarter view of BLACK MARKET BABE, #168. P-38J-5, 42-67147.

(Anderson)

(Tomberg)

#181 ROSE

(Krane)

(Anderson)

(Anderson)

(Anderson)

(Anderson)

P-38L-5, serial 44-25564

(Anderson)

475th Fighter Group Aerial Victories

Headquarters

Date	Time	Name	Type	Area
10/15/43	0840	C. MacDonald	2 Val / Val(Prob) / Val(Dam.)	Oro Bay
10/15/43	0840?	W.H. Ivey	2 Val	Oro Bay
10/23/43	1230	C. MacDonald	Oscar	Rabaul
10/25/43	1215	C. MacDonald	Zero	Rabaul
11/09/43	1020-30 / 1115	C. MacDonald	2(Oscar) / Oscar (Prob)	Alexishafen
11/16/43	0945	M.M. Smith	2 (Oscar)	Wewak
12/21/43	1200	C. MacDonald	2 Val	Arawe, N.B.
12/22/43	0950	M.M. Smith	Tony	Near Borum
01/10/44	1105	C. MacDonald	Tony	Wewak
01/18/44	1050	C. MacDonald	(Oscar)	Wewak
02/03/44	1155	M.M. Smith	Lily	Near But
06/08/44	1345	C. MacDonald	Zero	60 mi. NW Manokwari
06/16/44	1245	M.M. Smith	Val	Near Jefman I.
08/01/44	1235 / 1245	C. MacDonald	Rufe / Val	10 mi. S. of Koror I. / Koror I.
08/01/44	1240	M.M. Smith	Rufe	13 mi. S. of Koror I.
08/01/44	1240	D.P. Miller, Jr.	Hamp(prob.)	Koror I.
10/14/44	1045	M.M. Smith	Hamp	Balikpapan
11/10/44	0905	C. MacDonald	Oscar	Ormoc Bay, Leyte
11/11/44	1245	C. MacDonald	2 Jack	off Apalo pt., Ormoc
11/28/44	0745	C. MacDonald	Zero	Ormoc Bay
12/07/44	1125 / 1430	C. MacDonald	Jack / 2 Jack	S. of Ormoc Bay / SW end of Negros
12/07/44	1150	M.M. Smith	2 Jack	Ormoc Bay
12/13/44	1740	C. MacDonald	Sally	SW coast of Negros
12/25/44	1045-1145	C. MacDonald	2 Jack / Zero / Jack(prob.)	Clark Field
01/01/45	1055	C. MacDonald	Dinah / Tojo	Clark Field / 8-10 miles E. of Clark
02/13/45	1155	C. MacDonald		150 miles east of Indo-China
03/28/45	1155	J. Loisel	Frank	S. Coast of Tre I.

431st Fighter Squadron

Date	Time	Name	Type	Area
08/16/43	1520	D.W. Allen	Oscar / Kate	Marilinan
08/16/43	1520	H.W. Brown	3 (Oscar)	Marilinan
08/16/43	1520	W.R. Lewis	Oscar	3 mi. NE of Tsili Tsil
08/16/43	1520	L.C. Lutton	Oscar	Marilinan
08/16/43	1520	J.C. Mankin	2 (Oscar)	Marilinan
08/16/43	1520	P. Smith	Kate	Marilinan
08/16/43	1520	A.E. Wenige	2 (Oscar)	Marilinan
08/18/43	0945	D.C. Bellows	(Oscar)	Near Wewak
08/18/43	0945	R.F. Cline	(Oscar)	Near Wewak
08/18/43	0945	E.J. Czarnecki	2 (Oscar)	Near Wewak
08/18/43	0945	V.E. Jett	2(Oscar)	Near Wewak
08/18/43	0945	F.J. Lent	(Oscar)	Near Wewak
08/18/43	0945	L.C. Lutton	2 Oscar	Near Wewak
08/18/43	0945	T.B. McGuire	2 (Oscar) / Tony	Near Wewak
08/18/43	0945	W.G. Sieber	(Oscar)(prob.)	Near Wewak
08/18/43	0945	W.F. Haning	(Oscar)(prob.)	Near Wewak
08/21/43	1005	D.W. Allen	3 (Oscar)	Near Wewak
08/21/43	1005	E.J. Czarnecki	Nate (Oscar?)	Near Wewak
08/21/43	1005	R.E. Dunlap	(Oscar)	Near Wewak
08/21/43	1005	V.E. Jett	(Nick)	Near Wewak
08/21/43	1005	F.J. Lent	(Nick, Oscar)	Near Wewak
08/21/43	1005	T.B. McGuire	2 (Oscar) / (Nick) (Dam.)	Near Wewak
08/29/43	1005	F.A. Nichols	Oscar	Near Wewak
08/29/43	1050	H.E. Holze	(Oscar)	Near Wewak
08/29/43	1050	T.B. McGuire	(Oscar) Tony	Near Wewak
09/13/43	1025	V.T. Elliott	(Oscar)	Near Wewak
09/13/43	1025	V.E. Jett	(Nick)	Near Wewak
09/28/43	1025	F.F. Champlin	(Oscar)	Near Wewak
09/28/43	1145	J.J. Hood	(Oscar)Tony	Wewak
09/28/43	1145	W.R. Lewis	(Oscar)	Near Wewak
09/28/43	1145	T.B. McGuire	2 (Oscar)	Near Wewak
10/15/43	0840	E.J. Czarnecki	Zero (Zero prob.)	Oro Bay
10/15/43	0840	V.T. Elliott	Val	Oro Bay
10/15/43	0840	W.C. Gronemeyer	Zero	Oro Bay
10/15/43	0840	M.F. Kirby	Val	Oro Bay
10/15/43	0840	F.J. Lent	2 Zero, Val	Oro Bay
10/15/43	0840	T.B. McGuire	Va1(2 Zero prob.)	Oro Bay
10/15/43	0840	F.H. Monk	Zero	Oro Bay
10/15/43	0840	P.V. Morriss	Zero	Oro Bay
10/15/43	0840	W.G. Sieber	Val	Oro Bay
10/15/43	0840	R.E. Hunt	Zero Prob.	Oro Bay
10/17/43	1015	E.J. Czarnecki	2 Zero	Near Buna Bay
10/17/43	1015	V.T. Elliott	Zero	Near Buna Bay
10/17/43	1015	C.H. Houseworth	2 Zero / Zero Prob.	Near Buna Bay
10/17/43	1015	M.F. Kirby	Zero	Near Buna Bay
10/17/43	1015	T.B. McGuire	3 Zero	Buna Bay
10/17/43	1015	R.E. Hunt	Zero(prob)	Buna Bay
10/17/43	1015	K.M. Richardson	Zero(prob)	Buna Bay
10/23/43	1215	V.T. Elliott	2 Zero	Rabaul
10/23/43	1215	M.F. Kirby	Hamp	Over Rabaul
10/23/43	1215	L.C. Lutton	Zero	Near Rabaul
10/23/43	1215	A.E. Wenige	Hamp(prob)	Rabaul
10/24/43	1110	H.W. Brown	Zero	Near Rabaul
10/24/43	1110	F.J. Lent	Tony	Over Rabaul
10/24/43	1110	W.R. Lewis	Zero(prob)	Near Rabaul
11/02/43	1340	F.F. Champlin	2 Zero	Near Rabaul
11/02/43	1340	R.E. Hunt	2 Zero	Near Rabaul
11/02/43	1340	M.F. Kirby	2 Zero	Rabaul
11/02/43	1340	F.J. Lent	Zero(Zero Prob)	Near Rabaul
11/02/43	1340	L.C. Lutton	Zero	Near Rabaul
11/02/43	1340	F.H. Monk	Zero	Near Rabaul
11/02/43	1340	A.E. Wenige	2 Zero	Near Rabaul
11/07/43	1230	R.L. Herman	Zero	Rabaul
11/07/43	1230	H.E. Holze	Tony	Rabaul
11/07/43	1230	J.C. Mankin	Zeke(Prob) / Tony, Oscar	Rabaul
11/07/43	1230	A.E. Wenige	Tony	Rabaul
12/16/43	1355	D.W. Allen	(Helen)	Near Arawe
12/16/43	1355	F.J. Lent	(Helen)	Near Arawe
12/16/43	1355	J.A. Tilley	(Helen)	Near Arawe
12/16/43	1355	T.B. McGuire	Zero Prob.	Near Arawe
12/22/43	0950	D.C. Bellows	(Oscar)	Near Boram
12/22/43	0950	W.S. O'Brien	(Oscar)	Near Boram
12/22/43	0950	V.T. Elliott	Tony Dam.	Boram
12/22/43	0950	A.R. Kidd	(Oscar Dam)	Boram
12/26/43	1435	F.F. Champlin	Val	Cape Gloucester
12/26/43	1435	V.T. Elliott	Zero, Oscar	Cape Gloucester
12/26/43	1435	T.B. McGuire	3 Val	Cape Gloucester
12/26/43	1435	F.H. Monk	Zero	Cape Gloucester
12/26/43	1435	P.V. Morriss	Val	Cape Gloucester
12/26/43	1435	O.E. Powell	Val	Cape Gloucester
12/26/43	1435	H.E. Zehring	2 Val	Cape Gloucester
12/26/43	1435	J.A. Cohn	2 Val prob.	Cape Gloucester
12/26/43	1435	V.E. Jett	Val Dam.	Cape Gloucester
01/18/44	1050	V.E. Jett	(Oscar)	Wewak
01/18/44	1050	J.A. Tilley	Oscar Dam.	Wewak
02/03/44	1155	W.D. Clodfelter	Lily	NE But
02/03/44	1155	J.A. Cohn	Lily	NE But
02/03/44	1155	F.H. Fulkerson	Lily	But
02/03/44	1155	W.C. Gronemeyer	Lily	NE But
02/03/44	1155	V.E. Jett	Lily	North of But
03/30/44	1037	A.R. Kidd	Tony Dam.	Hollandia
03/31/44	1015	R.L. Crosswait	(Oscar)	S. of Hollandia
03/31/44	1015	A.R. Kidd	(Oscar)	S. of Hollandia

Date	Time	Name	Type	Area
03/31/44	1015	F.J. Lent:	2(Oscar)	S. of Hollandia
03/31/44	1015	F.H. Monk	(Oscar)	Hollandia
03/31/44	1015	H.W. Zehring	2(Oscar)	Hollandia
04/03/44	1200	J.A. Cohn	Oscar	S, of Ebeli Plant., Hollandia
05/17/44	1210	T.B. McGuire	Oscar	Noemfoor I.
05/17/44	1210	W.S. O'Brien	Oscar	Noemfoor I.
05/17/44	1210	J.A. Tilley	Oscar	Noemfoor I.
05/19/44	1340	T.B. McGuire	Tojo	Manokwari, N.G.
06/04/44	1040	P.V. Morriss	Oscar	Near Mois Neom I.
06/16/44	1245	W.C. Gronemeyer	Sonia	Near Jefman I.
06/16/44	1245	T.B. McGuire	Sonia, Oscar	Near Jefman I.
06/16/44	1245	F.H. Monk	Oscar	Near Jefman I.
06/16/44	1245	P.V. Morriss	2 Oscar	Near Jefman I.
06/16/44	1245	W.S. O'Brien	Oscar	Near Jefman I.
06/16/44	1245	E.Provencio	Oscar	Near Jefman I.
06/16/44	1245	H.B. Reeves	Oscar	Near Jefman I.
07/27/44	1300	G.M. Veit	Oscar	Lolobata, Halmaheras
07/27/44	1300	F.J. Benevent	Oscar	Lolobata, Hamahera
07/27/44	1300	T.B. McGuire	Oscar	Lolobata
08/04/44	1145	W.S. O'Brien	Tony	Amboina
10/14/44*	1050	T.B. McGuire	Oscar, Hamp, Tojo, Oscar Prob.	Balikpapan
11/01/44	1145	T.B. McGuire	Tojo	San Pablo, Leyte, P.I.
11/10/44	1445	T.B. McGuire	Oscar	30 mi. NW of Ormoc
11/12/44	0815	F.F. Champlin	Oscar,Lily	30 mi NNE of Dulag
11/12/44	0815	J.A. Moering	Oscar	NE Leyte Gulf
11/12/44	1710	W.T. Hudnell, Jr.Jack		Northern Cebu
11/12/44	1710	R.F. Cline	Jack	Northern Cebu
11/12/44	1710	T.B. McGuire	2 Jack	Northern Cebu
11/12/44	1825	K.F. Hart	Oscar Prob.	Dulag Harbor
11/18/44	0725	G.G. Dewey	Oscar	Tacloban
11/18/44	0725	W.F. Ekdahl	Oscar	Tacloban
11/18/44	0730	J.A. Moering	Oscar	Tacloban
11/24/44	0810	J.A. Barnes	Tony	Carigara, Leyte
11/24/44	0810	K.F. Hart	2 Tony	Carigara, Leyte
11/24/44	0810	R.A. Koeck	2 Oscar	Carigara, Leyte
11/24/44	0810	J. Pietz, Jr.	2 Oscar, Kate	Leyte
11/24/44	1640	T.M. Oxford	Kate	Near Dulag
11/24/44	1755	K.F. Hart	Jake	SW of Carigara
11/27/44	1155	L.D. DuMontier	Jack	Near Ormoc
12/02/44	1707	K.F. Hart	Val	Near Ormoc
12/05/44	1725	M. Olster	Oscar	Cebu, Cebu I.
12/06/44	1415	W.C. Gronemeyer	Lily	Dinagat Sound
12/06/44	1415	C.C. Parshall	Lily	Dinagat I.
12/07/44	1145	T.B. McGuire	Oscar	Poro I.
12/07/44	1610		Tojo	Near Ormoc Bay
12/07/44		J.B. Rittmayer	Kate, Tojo	Ormoc Bay
12/07/44	1610	F.H. Fulkerson	Tojo	Near Ormoc Bay
12/07/44	1615	F.F. Champlin	Zero	Maripipi I.
12/07/44	1615-1810	K.F. Hart	2 Oscar	E of Olango I.,Camotes Sea
12/07/44	1630	J.B. Rittmayer	Tojo	Ormoc Bay
12/07/44	1640	W.F. Ekdahl	Hamp	Cebu I.
12/07/44	1700	D.S. Thropp. Jr.	Hamp	Damulog, Cebu
12/07/44	1805	E. Provencio	Zero	Camotes I.
12/11/44	1700	J.A. Tilley	Zero	NW of Ormoc Bay
12/12/44	1710	H.B. Reeves	Zero	Mountain on S. tip of Leyte
12/12/44	1710	A.R. Neal	Zero	S. Tip of Leyte
12/13/44	1710	T.B. McGuire	Jack	Tanza, Negros
12/17/44	1620	J.B. Rittmayer	Oscar	San Jose, Mindoro
12/25/44	1035	J.R. Ballard	Zero	Near Clark Field
12/25/44	1035	F.F. Champlin	Oscar	Near Clark Field
12/25/44	1035	L.D. DuMontier	2 Zero	NW of Clark
12/25/44	1035	H.W. Grady	Jack	Near Clark
12/25/44	1035	T.E. Martin	Zero	Near Clark
12/25/44	1035-1045	T. B. McGuire	3 Zero	Clark area
12/25/44	1035	J.A. Moering	Zero	NE of Clark
12/25/44	1035	A.R. Neal	Zero	Near Santa Rosa
12/25/44	1035	J.A. Tilley	Zero	E. of Mabalacat
12/25/44	1040-1045	F.H. Fulkerson	2 Jack	Near Clark
12/25/44	1040-1045	J. Pietz, Jr.	2 Jack	Near Clark
12/25/44	1055	T.M. Oxford	Jack, Zero	Near Clark
12/26/44	1040-1050	T.B. McGuire	4 Zero 52	Near Clark
12/26/44	1045	F.F. Champlin	Zero	Near Clark
12/26/44	1050	H.B. Reeves	Zero	S. Clark Field
12/26/44	1100	J.A. Tilley	Zero 52	NW of Clark
01/01/45	1120	C.J. Herman	Zero 52	Near Sula, Luzon

Date	Time	Name	Type	Area
01/04/45	1150	H.B. Reeves	Zero	Near Cabangan, Luzon
02/25/45	1140	J.H. Barnes	Rufe	Cam Ranh Bay, lndo-China
02/25/45	1140	T.E. Martin	Rufe	Cam Ranh Bay, lndo-China
03/28/45	1205	J. Pietz	Tojo	Ben Goi Bay, lndo-China
03/28/45	1205	K.F. Hart	2 Hamp	Ben Goi Bay, lndo-China
03/28/45	1205	T.E. Martin	Hamp	S. Ben Goi Bay
03/28/45	1205	H.B. Reeves	2 Hamp	Ben-Goi Bay
03/29/45	1135	J.R. O'Rourke	Zero 52	Cape Batangan, lndo-China
			Zero 52 Dam.	

432nd Fighter Squadron

Date	Time	Name	Type	Area
08/18/43	0935	W.M. Waldman	(Oscar) (Oscar da,.)	Near But, N.G.
08/18/43	0935	R.S. Ryrholm	(Oscar) Dam.	Wewak
08/18/43	0935	V.A. Cloud	(Oscar Prob.)	Dagua
08/21/43	0935	E. Summer	(Oscar)	Near Dagua
08/21/43	0940	B.M. Gresham	(Nick)	Near Dagua
08/21/43	0945	J.R. Farris	(Nick prob)	Wewak
08/21/43	0945	G.D. Gholson	(Nick)	Near Dagua
08/21/43	0945	F.A. Harris	2Tony 4 Tony Dam.	Near Dagua
08/21/43	0945	J.S. Loisel	2 Tony	Near Dagua
08/21/43	0945	P.W. Lucas	(Oscar)	Near Dagua
08/21/43	0945	D.T. Roberts	2 (Oscar)	Near Dagua
08/21/43	0945	C.P. M. Wilson	2 TE fighter	S of Dagua
08/21/43	0945	J.E. Michener	(Oscar prob)	Near Dagua
08/21/43	0945	T.J. Simms	Tony prob.	Wewak
09/07/43	1320	C.O. Bartlett	Oscar	Near Morobe Harbor
09/07/43	1320	N.R. Lundy	(Oscar)	Morobe Harbor
09/07/43	1320	D.T. Roberts	Oscar	Near Morobe Harbor
09/07/43	1320	T.J. Simms	Oscar	Near Morobe Harbor
09/13/43	1050	N.R. Lundy	Tony	Angoram, N.C.
09/13/43	1050	F.D. lomkins	(Oscar)	30 mi. W. of Nubia, N.G.
09/20/43	1140	B.M. Gresham	2 Tony	Over Marilinan
09/20/43	1250	L.C. Bryan	(Oscar)	Over Marilinan
09/20/43	1130	W.M. Waldman	Tony prob.	Over Marilinan
09/22/43	1240	J.R. Farris	Betty	25 mi. off Finschhafen
09/22/43	1245	J.C. Ince	Betty, Zero	25 mi. off Finschhafen
09/22/43	1245	C.J. Ratajski	Zero	25 mi. off Finschhafen
09/22/43	1245	J.T. Rundell	Betty	25 mi. off Finschhafen
09/22/43	1245	C.B. Willis	Betty	25 mi. off Finschhafen
09/22/43	1250	L.C. Bryan	2 Zero	Over Finschhafen
09/22/43	1250	V.A. Cloud	Betty (Zero)	25 mi. off Finschhafen
09/22/43	1250	H.L. Condon	Zero	Over Finschhafen
09/22/43	1250	Z.W. Dean	Betty, Zero	25 mi. off Finschhafen
09/22/43	1250	F.A. Harris	2 Zero, Betty	25 mi. off Finschhafen
09/22/43	1250	H.A. Hedrick	Zero	25 mi. off Finschhafen
09/22/43	1250	J.S. Loisel	Zero	25 mi. off Finschhafen
09/22/43	1250	J.E. Michener	Zero Prob.	25 mi. off Finschhafen
09/30/43	0845	H.L. Condon	Betty	Finschhafen area
10/12/43	1050	J.E. Fogarty	Betty	5 mi. S. of Rabaul*
10/15/43	0805	J.R. Farris	Zero	Oro Bay
10/15/43	0810	J.S. Loisel	2 Zero	Oro Bay
10/15/43	0825	E. Summer	Oscar	Oro Bay
10/15/43	0830	Z.W. Dean	Val	Oro Bay
10/15/43	0830	B.M. Gresham	Nate (Val?)	Oro Bay
10/15/43	0830	L.M. Mayo	Oscar	Oro Bay
10/15/43	0830	F. D. Tomkins	Zero, Val	Off Oro Bay
10/15/43	0835	V.A. Cloud	Val	Oro Bay
10/15/43	0835	F.A. Harris	2 Val	Oro Bay
10/23/43	1235	H.A. Hedrick	(Zero)	Near Rabaul
10/23/43	1235	C.P.M. Wilson	(Zero)	Near Rabaul
10/23/43	1225	J.F. Weber	Oscar Prob.	Rabaul
10/23/43	1230	E.G. Dickey	Oscar Prob.	Rabaul
10/24/43	1150	B.M. Gresham	Zero	Rabaul
10/24/43	1150	F.A. Harris	Zero, Zero Prob.	Rabaul
10/24/43	1155	Z.W. Dean	Zero, Hamp	Rabaul
10/24/43	1210	G.D. Gholson	Oscar(Zero?)	SW of Tobera
10/24/43	1210	E. Summer	Oscar(Zero?)	Rabaul
10/24/43	1210	C.J. Mann	Zero Prob.	Rabaul
11/02/43	1330	G.D. Gholson	Zero, Oscar	Rabaul

* McGuire's three victories of 10/14/44 are carried on the rolls of both the 9th and 431st Fighter Squadrons.

* Scored while flying with 1st Provisional Squadron, but added to 432nd total.

Date	Time	Name	Type	Area
11/02/43	1335	A.L. Peregoy	Zero	Rabaul
11/02/43	1335	F.E. Hanson	Oscar Prob.	Rabaul
11/02/43	1340	H.A. Hedrick	Oscar, Zero Prob.	Rabaul
11/02/43	1340	L.M. Mayo	Tony	Rabaul
11/02/43	1340	T. Fostakowski	(Prob.)	Rabaul
11/09/43	1015	P.J. Dahl	(Oscar)	Alexishafen
11/09/43	1015	J.C. Ince	(Oscar)	Alexishafen
11/09/43	1015	J.T. Rundell, Jr.	2 (Oscar)	Alexishafen
11/09/43	1015	C.B. Willis	(Oscar)Prob.	Alexishafen
11/09/43	1020	J.M. Forster	(Oscar)prob.	Alexishafen
11/16/43	0950	J.C. Ince	(Oscar)	Wewak
11/16/43	0950	J.M. Forster	(Oscar)prob.	Wewak
12/13/43	1110	J.S. Loisel	Zero	Gasmata
12/21/43	1200	J.R. Farris	Val	Arawe
12/21/43	1200	J.S. Loisel	Zero	Arawe
12/21/43	1200	J.E. Michener	Oscar	Arawe
12/21/43	1200	A.L. Peregoy	Val	Arawe
12/21/43	1200	W.T. Ritter	2 Val	Arawe
12/21/43	1200	E. Summer	2 Vall	Arawe
12/22/43	1000	V.A. Cloud	Tony(Oscar)	Wewak
12/22/43	1000	P.J. Dahl	(Oscar)	Wewak
12/22/43	1000	Z.W. Dean	2 (Oscar)	Wewak
12/22/43	1000	E. Summer	(Oscar)	Wewak
12/23/43	0930	C.J. Ratajski	Sally	(Muroc Lake?)
01/18/44	1050	B.M. Gresham	(Oscar)	Wewak
01/18/44	1050	R.L. Hadley	(Oscar)	Wewak
01/18/44	1050	H.A. Hedrick	(Oscar)	Wewak
01/18/44	1055	W.T. Ritter	Tony	Wewak
01/23/44	1110	P.J. Dahl	(Oscar)	Wewak
01/23/44	1110	J.S. Loisel	(Oscar)	Brandi Plaat., near Wewak
04/03/44	1135	H.L. Condon	Oscar	S. side Sentani Lake, Hollandia
04/03/44	1135	P.J. Dahl	2(Oscar)	Sentani Lake, Hollandia
04/03/44	1135	J.M. Forster	3 Oscar	Hollandia
04/03/44	1135	J.L. Hannan	Oscar	Hollandia
04/03/44	1135	J.S. Loisel	2 (Oscar)	Hollandia
04/03/44	1135	E. Summer	(Oscar)	Hollandia
04/03/44	1135	J.W. Temple	(Oscar)	Hollandia
04/03/44	1135	J.E. Michener	(Oscar) Dam.	Hollandia
06/08/44	1335	P.J. Dahl	Oscar, Oscar Dam.	N. of Cape Waios
06/08/44	1335	C.J. Mann	Oscar	N. of Cape Waios
06/16/44	1245	R.L. Hadley	2(Oscar)	Jefman
10/14/44	1045	J.M. Forster	Zero	Balikpapan
11/02/44	1045	J.M. Forster	Oscar	Ponson I., W. Ormoc Bay
11/08/44	0610	J.M. Forster	Oscar	Dulag, Leyte
11/10/44	0905	R.D. Kiick	2 Tony	NW of Ponson I.
11/10/44	0905	H.C. Toll	2 Tony	NW of Ponson I.
11/10/44	0910	P.J. Dahl	Tony	NW of Ponson I.
11/10/44	0910	D.W. Olson	Tony	NW of Ponson I.
11/10/44	0910	C.J. Ratajski	Tony	NW of Ponson I.
11/10/44	0910	R.B. Schuh	Tony	NW of Ponson I.
11/10/44	0910	D.W. Willis	Tony	Ormoc
11/10/44	1600	L.C. LeBaron	Tony	Ormoc
11/10/44	1600	L.D. Roberts	Tony	Ormoc
11/11/44	1015	J.L. Hannan	Oscar	Ormoc Bay
11/11/14	1245	G.R. Beatty	2 Jack	Ormoc Bay
11/12/44	1330	E. Summer	2 Zero	SW of Tacloban
11/12/44	1730	C.O. Anderson	Oscar, Hamp	Dulag – N. Cebu
11/12/44	1330	H.D. Owen	Zero Dam.	Tacloban
	1820		Zero	Dulag
11/24/44	0800-0820	P.W. Lucas	2 Tony	20 mi. from Tacloban
11/24/44	1115-1730	R.H. Kimball	Hamp, Kate	Baybay, Leyte
11/27/44	1055	P.W. Lucas	2 Zero	SW of Dulag
11/27/44	1055	L.L. Parlett	Zero	Leyte Gulf
11/27/44	1130	G.R. Beatty	Jill	Leyte Gulf
11/27/44	1130	L.A. Dowler	Zero	Leyte Gulf
11/27/44	1200	H.C. Tcll	Zero	Leyte Gulf
12/06/44	1355	C.J. Ratajski	Jack	Tanza, Negros
12/06/44	1630	A.P. Morris	Nick	Cebu
12/07/44	0750-1125	J.M. Forster	Zero, Dinah Dinah Prob.	Ormoc Bay W. Dulag
12/07/44	0940	E. Summer	Oscar	Ponson-Poro ls.
12/07/44	1115-1125	R.H. Kimball	(George?)	Ormoc Bay
12/07/44	1430	L.W. Blakely	Zero	Ormoc Bay
12/07/44	1440	A.A. McKenzie	Dinah	W. Cebu
12/07/44	1620	L.D. Roberts	(George?)	Ormoc Bay
12/11/44	1315	P.W Lucas	Zero Prob.	SW oi Masbate
12/25/44	1045	H.D. Owen	2 Zero	Mabalacat
12/25/44	1050	H.L. Condon	2 Zero	Mabalacat
12/25/44	1050	J.M. Forster	Zero, Zero Dam.	Mabalacat

Date	Time	Name	Type	Area
12/25/44	1055	L.J. Noblitt	Zero	Mabalacat
01/01/45	1110	p.W. Lucas	Zero	Clark Field
03/05/45	1150	P.J. Dahl	Sally	Taiko, Formosa
03/28/45	1150	P.J. Dahl	Hamp	Ben Goi Bay
03/28/45	1200	C. Wacker	Hamp	N. Nha Trang
03/29/45	1135-1155	L.A. Dowler	2 Zero	Cape Batangan
03/29/45	1135-1155	L.C. LeBaron	3 Zero	Cape Batangan
03/29/45	1140	H.D. Owen	Zero	Cape Batangan

433rd Fighter Squadron

Date	Time	Name	Type	Area
09/02/43	1000	C.C. Wire	(Oscar) (Osc. prob.)	Wewak
09/02/43	1000	J.C. Smith	(Oscar) Dinah Prob.	Wewak
09/04/43	1000	R.T. Cleage	(Oscar) Prob.	Wewak
09/04/43	1400	H.H. Jordan	Betty	SE oi Salamaua
09/04/43	1400	J.T. McKeon	Oscar	SE of Salamaua
09/04/43	1400	J.K. Parkansky	Oscar	SE of Salamaua
09/04/43	1405	K.D. Kirschner	Betty	SE of Salamaua
09/04/43	1400	D.O. Meyer	(Oscar)Prob.	SE of Salamaua
09/22/43	1000	J.T. McKeon	Tony	25 mi. S. of Finschhaien
09/24/43	1230	J.T. McKeon	Hamp, Hamp Dam.	Finschhafen
09/24/43	1230	R.T. Cleage	Hamp Prob.	NW of Finschhafen
09/26/43	1010	C.C. Wire	2 (Oscar)	Wewak
09/26/43	1010	C.M. Wilmarth	(Oscar, TonyProb.)	Wewak
09/26/43	1010	J.S. Babel	Tony Prob.	Wewak
09/26/43	1010	M.L. Low	Tony Prob.	Wewak
10/12/43	1040	R.T. Cleage	Oscar	Rabaul
10/15/43	0845	R.T. Cleage	Oscar(Zero?)	Oro Bay
10/15/43	0845	V.A. Hagan	Oscar(Zero?)	Oro Bay
10/15/43	0845	W.G. Jeakle	Val	Oro Bay
10/15/43	0845	R.D. Kimball	Val	Oro Bay
10/15/43	0845	D.Y. King	Val	Oro Bay
10/15/43	0845	J.M. Malloy	Val, Val Prob.	Oro Bay
10/15/43	0845	D.D. Revenaugh	Val	Oro Bay
10/15/43	0845	J.C. Smith	Oscar(Zero?)	Oro Bay
10/17/43	1030	J.A. Fisk	2 Zero	Morobe Harbor
10/17/43	1030	C.A. Grice	Zero	Morobe Harbor
10/17/43	1030	J.H. Howard	Zero	Morobe Harbor
10/17/43	1030	D.O. Meyer	Zero	Morobe Harbor
10/17/43	1030	D.T. Roberts, Jr.	2 Zero	Morobe Harbor
10/17/43	1030	J.C. Smith	2 Zero	Morobe Harbor
10/17/43	1030	R.M. Tomberg	Zero	Morobe Harbor
10/23/43	1215	D.T. Roberts, Jr.	2 Zero	Rabaul
10/23/43	1215	D.O. Meyer	Zero	Rabaul
10/23/43	1230	J.C. Smith	Oscar(Zero?)	Rabaul
10/24/43	1110	J.M. Malloy	Zero, Zero Prob.	Rabaul
10/24/43	1115	C.C. Wire	2 Zero	Rabaul
10/24/43	1115	J.S. Babel	Zero, Zero Prob.	Rabaul
10/24/43	1115	H.W. Cochran	Zero	Rabaul
10/24/43	1115	J.A. Fisk	Zero	Rabaul
10/24/43	1115	D.Y. King	2 (Zero)	Rabaul
10/24/43	1115	D.T. Roberts, Jr.	Zero	Rabaul
10/24/43	1120	C.A. Grice	Zero	Rabaul
10/29/43	1230	C. Brenizer, Jr.	Zero	Rabaul
10/29/43	1230	H.W. Cochran	Zero	Rabaul
10/29/43	1230	C.A. Grice	Zero	Rabaul
10/29/43	1230	D.Y. King	Zero	Rabaul
10/29/43	1230	D.T. Roberts, Jr.	Zero	Rabaul
10/29/43	1230	J.C. Smith	Zero	Rabaul
11/02/43	1340	D.T, Roberts, Jr.	Zero*	Rabaul
11/02/43	1340	D.D. Revenaugh	Zero*	Rabaul
11/09/43	1025	J.A. Fisk	(Oscar)	Alexishafen
11/09/43	1030	J.S. Babel	(Oscar)	Alexishafen
11/09/43	1030	D.T. Roberts, Jr.	(Oscar)	Alexishafen
11/09/43	1030	W.R. Lewis	TonyProb., (Oscar) Prob., Dam.	Alexishafen
11/09/43	1030	W.J. Grady	(Oscar) Prob.	Alexishafen
11/09/43	1030	H.W. Cochran	(Oscar) Prob.	Alexishafen
11/16/43	1040	W.J. Grady	(Oscar)	Wewak
11/16/43	1040	W.G. Jeakle	Tony	Finschhafen
12/15/43	1115	C. Brenizer	Tojo?	Central New Britain
12/16/43	1400	C.J. Rieman	Tony	W. of Talasea
12/16/43	1400	W.G. Jeakle	Zero	Arawe
12/16/43	1400	W.R. Lewis	2 (Helen)	Arawe

*Originally listed as probables in most accounts, these two victories are now apparently widely accepted.

Date	Time	Name	Type	Area	Date	Time	Name	Type	Area
12/16/43	1400	J.S. Babel	(Helen)	Arawe	12/06/44	0845	G.J. Gabik	Jack	SW coast Panaon I.
12/16/43	1400	C.A. Grice	(Helen)	Arawe	12/06/44	0845	W.J. Hulett	Jack	SW Coast Panaon I.
12/18/43	1145	W.J. Grady	Zero	SE Cape Raoult, NB	12/06/44	0845	M.B. Ryerson	Jack Prob.	Ponson-Limasawa ls.
12/18/43	1145	S.C. Northrup	Zero	SE Cape Raoult, NB	12/07/44	1655	S.M. Morrison	Oscar	Ormoc Bay
12/18/43	1145	R.M. Tomberg	Tony	SE Cape Raoult, NB	12/07/44	1750	C. Brenizer, Jr.	Helen	Canigao Channel
12/18/43	1145	W.R. Lewis	Zero Dam.	SE Cape Raoult, NB	12/11/44	0900-0920	J.E. Purdy	2 Oscar, Oscar Dam.	N. Bantayan I.
12/22/43	0950	J.M. Malloy	(Oscar) (Sonia?)	Wewak	12/11/44	0915	W.J. Hulett	Oscar	N. Bantayan I.
03/15/44	1020	W.J. Grady	(Oscar)	Wewak	12/11/44	0915	J.E. Miller, Jr.	Oscar	NW coast Bantayan
03/15/44	1020	R.D. Kimball	Tony	Wewak	12/11/44	0915	C.R. Redding	Oscar	NW Bantayan
03/15/44	1020	P.R. Peters	Tony	Wewak	12/11/44	1300	S.M. Morrison	Zero	NW Ormoc
04/03/44	1145	W.R. Lewis	(Oscar)	Hollandia	12/14/44	0925	C.L. Ettien, Jr.	Tojo	E. Canlaon Volcano, Negros
04/03/44	1200	R.D. Kimball	Oscar	Hollandia	12/17/44	1628	J.E. Purdy	2 Zero 52	E. of Tombanac, Mindoro
05/16/44	1415	W.R. Lewis	Oscar	Noemfoor					
05/16/44	1415	J.E. Purdy	Oscar	Noemfoor	12/17/44	1618	W.J. Grady	Jack	NW of Ambulong I.
05/16/44	1415	H.V. Stiles	Oscar	Noemfoor	12/18/44	1300	J.A. Fisk	Jack, Zero	SW Mindoro
05/19/44	1000	W.R. Lewis	Pete	Noemfoor	12/18/44	1305	G.J. Gabik	Tony	S Mindoro
06/06/44	1210	P.R. Peters	Oscar	5 mi. S. of Babo A/D	12/18/44	1315	S.M. Morrison	Oscar, Oscar Prob.	Mangarin, Mindoro
06/16/44	1245	J.P. Price, Jr.	Oscar	Jefman					
06/16/44	1250	C.J. Rieman	Oscar	Jefman	12/18/44	1310	J.E. Sperling	Zero	E. of San Jose, Min.
06/16/44	1250	E.B. Roberts	Tojo	Jefman	12/25/44	1123	G.M. Maxwell	Zero 52	SE of Mabalacat
06/16/44	1250	W.R. Lewis	Ostar Dam.	Jefman-Samate	12/26/44	1048	J.A. Fisk	Tojo	SE runway 5, Clark Fld.
07/28/44	1045	C.A. Lindbergh	Sonia*	3 mi W. of Amahai	01/01/45	1105	E.B. Roberts	Jack	NE of Clark Field
11/19/44	1615	C.C. Wire	2 Oscar	W. coast of Leyte	01/01/45	1105	C.P.M. Wilson	Jack	
11/19/44	1645	P.C. Schoener	Val	S. of Dulag, Leyte				Jack Prob.?	Clark Field Area
11/24/44	0800-1445	J.A. Berry	Tony, (George)	Ormoc Bay				Jack Dam.?	
11/24/44	0800-1500	L.H. Ross	Tony, (George)	Ormoc Bay				Nick	E. of Manila
11/24/44	0810	C.A. Redding	Zero	Ormoc Bay	01/01/45	1105	J.R. Hammond	Jack	NE of Clark Field
11/24/44	1450	M.B. Ryerson	(George)	E. of Ponson I.	01/24/45	1810	L.H. Ross	Zero 52	Plaridel, Mindoro
11/24/44	1500	E.B. Edesberg	(George)	Ormoc Bay					
11/28/44	1648	W.L. Richmond, Jr.	Oscar	Ormoc Bay					
12/05/44	1710	J.E. Purdy	2 Val	Cebu Town, Cebu I.					
12/06/44	0845	A.L. Coburn	2 Jack, Jack Prob.	SW Coast Panaon I.					

* Victory not officially credited, but supported by sufficient evidence.

Aircraft of the 475th Fighter Group Aces

Colonel MacDonald by his original P-38 PUTT PUTT MARU – November-December 1943. (Campbell)

Wurtsmith, MacDonald and (probably) Freddie Smith with PUTT PUTT MARU in mid-1944. (Cooper)

2nd Putt Putt Maru in 1944. P-38J-15, serial 42-104024

Evidence of damage that 2nd Putt Putt Maru received in landing accident ca. July-August 1944. 3rd Putt Putt Maru went into service about the same time. (Krane files)

Left: 3rd Putt Putt Maru, P-38L-1, serial 44-24843, in December 1944. (Krane files)

24843 at about the end of 1944. (Carl Bong)

Dulag, Leyte. December 1944. 3rd Putt Putt Maru. (Krane files)

4th Putt Putt Maru, P-38L-5, serial 44-25643. This P-38 is listed as being lost on January 27, 1945 in a taxiing accident. 24843 is listed as being lost on January 2, 1945 with 432nd commander Henry Condon, which would give the 4th Putt Putt Maru a short operational life of less than a month. (Krane files)

Right side of 4th Putt Putt Maru nose with what seems to be the entire Headquarters maintenance crew. (475th history)

Left: 4th Putt Putt Maru with 26 victory marks. (H. Wolf)

Full left profile of 4th Putt Putt Maru. (Krane files)

Both sides of 5th Putt Putt Maru with final decorations.

Probably the 5th Putt Putt Maru, serial 44-25471 with original nose decoration application. (Krane files)

Final application of Putt Putt Maru markings. The number 100 was insignia blue with either a darker blue or black shadow border. (Krane files, except for A-16, which should be credited to George Jeschke)

Above and below: PUDGY II with 17 victories marked, probably late 1943 since McGuire tended to claim the fourth Val on December 26, 1943 even after it was granted to another pilot. Note the different texture and shade of the prop spinners. It is likely that at some point PUDGY II had at least one of the dark blue tipped spinners. (Krane collection)

Tom McGuire's PUDGY II with seven victories marked (August 29 to late September 1943). (Krane Collection)

PUDGY II undergoing engine change. (Krane collection)

PUDGY III, full left side view, 21 victory marks ca. June - July 1944

PUDGY III, probably May 1944. Note that R. Applewhite is still crewchief. (via D. Tabatt)

PUDGY III, 17 victories, Hollandia, May 1944 (Pete Madison)

PUDGY III, 21 victories, Hollandia or Biak. (Krane collection)

PUDGY III, McGuire with groundcrewmen Vander Geest and Kish.

Left: PUDGY III, with seventeen victory marks.

PUDGY III, full nose view with 21 victory marks. (Krane collection)

PUDGY III on Biak, mid-1944 (Krane files)

PUDGY III, scoreboard with 17 flags, May 1944. (via Crowe)

Full view of PUDGY IV with 22 victories on Biak.

PUDGY IV nose with 22 victories.

Full view of PUDGY V in October - November 1944.

Unusual angle of PUDGY V cockpit. Note boom stripes and shading of nose markings.

PUDGY V on Boroka strip, Biak.

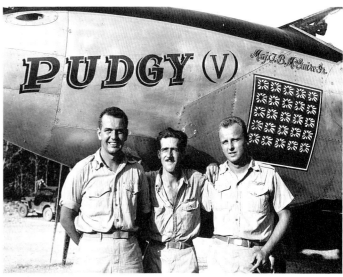

Closeup of PUDGY V nose, late October early November.

PUDGY V at 33 victory point, late December 1944. Since McGuire was never credited with exactly 33 victories, he either overestimated his own confirmed score or was awaiting confirmation of the 34th. (Cook)

Freshly enlarged scoreboard accommodates last score of 38 on PUDGY V.

Last markings of PUDGY V in December 1944 (Jeschke via Krane coll.)
Right side of PUDGY V with McGuire in the cockpit, Oct.-Nov. 1944.

Full left quarter of Colonel Loisel's P-38L-5 44-25443, Headquarters #101, April-May 1945.

Nose of Loisel's #101

Nose view of Loisel's #161 (Krane collection)

Meryl Smith's #101 with 6 victory markings in August-September 1944.

Vincent Elliott's #111 sometime in the early months of 1944. (Prell)

PEEWEE V in full right profile. (Krane files)

PEEWEE V, serial 44-25863 in full left profile. (Krane files)

Ken Hart's #111 PEEWEE V sometime in mid-1945. (Krane files)

Champlin's P-38J-10, serial 42-67597, #113, BUFFALO BLITZ.

Champlin's P-38H-1, serial 42-66558, left side: BUFFALO BLITZ; right side: "We Dood It", about the time of the November 2, 1943 Rabaul raid. (Krane coll.)

#124 left nose closeup. (Champlin)

Right: #124 in full left view. EILEEN inscribed on left engine cowling. (Champlin)

Champlin's #112 about the time that Hal Gray began flying it late in 1944. (Gray)

Closeup of #112, EILEEN. (Krane collection)

Captain Pietz's #126, VICKIE around the beginning of 1945.

#126, serial 44-25639, VICKIE, around April 1945. (Krane files)

#126, serial 44-25639, in flight before full markings were applied (Krane files)

Tilley's #116, P-38J-15, RANOOKI MARU, sometime in mid-1944 before Sgt Allan went home in October. (Tabatt)

Tilley's BETTY ANN in 1945. (Tilley)

Only known photo of Kirby's #125 MAIDEN HEAD HUNTER, #125. (Tabatt)

Frank Lent's #134, T-RIGOR MORTIS III, after he had completed his tour by mid-1944. (Gregg)

Nose of #134, serial 42-104032, with 49th FG pilot, John Bodak, standing by propeller. (Bodak)

TSgt Milton Back by Lent's #134. Back went on to crew the P-38 of another 431st ace, John "Rabbit" Pietz, later in 1944. (Krane files)

Elliot Summer's P-38L-5, 44-25600, late in 1945. (Tabatt)

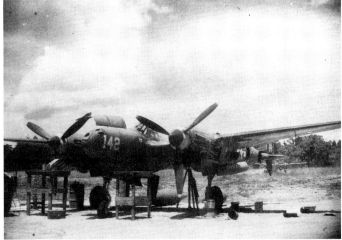

Fred Harris's #142, right side. (Krane files) #142, left side, undergoing maintenance around September 1943. (Krane Files)

Crewchief TSgt Rozell Stidd by #146, P-38J-15, serial 42-104454, pilot: Zach Dean. (Stidd)

Forward left quarter of Dean's #146 with groundcrew.

Groundcrewmen by #149 FLORIDA CRACKER, P-38L-5, serial 44-25132 Crewchief: TSgt Morrow. (Krane collection)

Right: #149, probably sometime in 1945. Crewchief was Joe Morrow and Asst. CC was Chester Rochon. (Krane collection)

Perry Dahl by his #162, 23 SKIDOO, P-38H-1, serial 42-66504 (Dahl)

Above and left: Billy Gresham's #168. (Krane collection)

Full front quarter of #197. (Hanks via Krane collection)

Nose of Danny Roberts's #197, P-38H-5, serial 42-66752. (Hanks via Krane collection)

Above and below: Left side of Warren Lewis's #170; (Anderson)

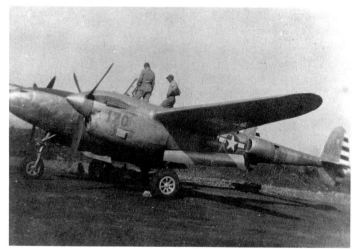

Right side of Lewis's #170, P-38J-15, serial 42-104305 (Anderson)

Lewis by his #170, 42-104305. (Krane collection)

Left front quarter of #170. This was one of the first natural metal P-38J aircraft to enter 433rd FS service in early 1944 and lasted until at least September 1944. (Krane collection)

Nose close-ups of Paul Morriss's #117. (John Campbell)

Wood Clodfelter's #124 "Pistol Packin' Ma-Ma." (Campbell)

Close-up of #151, Strictly Sex. (Campbell)

Nose of Calvin Wire's P-38L-5, serial 44-25880, LITTLE EVA, when Wire was commander of the 433rd, December 1944 through May 1945. TSgt George Rath was Crewchief and SSgt Smith was asst. CC. (Wire)

#114, Margie's Mare (?), apparently assigned to John Cohn early in 1944.

Jack Purdy's P-38L-5, serial 44-25930, listed in 433rd records as lost in a crashlanding on January 9, 1945. Purdy had two narrow escapes and isn't sure whether he lost LIZZIE V in the December 1944 crash or the one in January 1945. On one or the other he was flying the 433rd commander's aircraft. (Krane collection)

Bibliography

Government Publications

The Fifth Air Force in the Huon Peninsula Campaign, January to October 1943(AAFRH-13)

The Fifth Air Force in the Conquest of the Bismarck Archipelago, October 1943 to February 1944(AAFRH-16)

The Fifth Air Force in the Conquest of the Bismarck Archipelago, November 1943 to March 1944(AAFHS:43)

COMMAND DECISIONS, Edited by Kent Roberts Greenfield, Office of Chief of Military History, U.S. Army, 1960

USAAF FIGHTER OPERATIONS IN THE SOUTHWEST PACIFIC: ROLE OF THE 475th FIGHTER GROUP, Brammeier, Major Charles L., Air Command and Staff College Student Report 87-0320

Books

Steve Birdsall, FLYING BUCCANEERS: THE ILLUSTRATED STORY OF KENNEY'S FIFTH AIR FORCE, Doubleday, 1977

W.S. Churchill, THE SECOND WORLD WAR, VOL. VI, TRIUMPH AND TRAGEDY

Roy Cross, LOCKHEED P-38 LIGHTNING DESCRIBED, KOOKABURRA TECHNICAL PUBLICATIONS

Jeffrey L. Ethell, P-38 LIGHTNING AT WAR, Scribner's 1977 (co-author, Joe Christy)

– P-38 LIGHTNING, Crown 1984

Stanley Falk, LIBERATION OF THE PHILIPPINES, Ballantine Books 1971

Gene Gurney, P-38 LIGHTNING, ARCO 1969

Hata and Izawa, JAPANESE NAVAL ACES AND FIGHTER UNITS IN WWII, Naval Institute Press 1989 (Translated by Don Gorham)

W.N. Hess, PACIFIC SWEEP, Doubleday 1974

L.J. Hickey, WARPATH ACROSS THE PACIFIC, International Research and Publishing 1984

G.C. Kenney, GENERAL KENNEY REPORTS, Duell 1949

B.H. Liddell Hart, HISTORY OF THE SECOND WORLD WAR, G.P. Putnam's Sons 1972

E.T. Maloney, LOCKHEED P-38 LIGHTNING, Aero Publishers 1969

S.E. Morison, HISTORY OF UNITED STATES NAVAL OPERATIONS IN WORLD WAR II

– VOL.VI, BREAKING THE BISMARCKS BARRIER

– VOL.XII, LEYTE

Lex McAulay, INTO THE DRAGON'S JAWS: THE FIFTH AIR FORCE OVER RABAUL, Champlin Press 1986

E.R. McDowell, P-38 LIGHTNING, Arco-Aircam 1969

Kenn Rust, FIFTH AIR FORCE STORY, Historical Aviation Album 1973

John Stanaway, PETER THREE EIGHT, Pictorial Histories Pub. Co., 1986

John Vader, NEW GUINEA: THE TIDE IS STEMMED, Ballantine 1971

Ray Wagner, AMERICAN COMBAT PLANES, Doubleday 1968

LeRoy Webber, P-38J-M, Aircraft Profiles #106

Ronald Yoshino, LIGHTNING STRIKES, 1988 (excerpts)

Magazine Articles

Carroll R. Anderson
– "The Gentle Ace," Aerospace Historian
– "Black Sunday"
– "McGuire's Last Mission," Air Force, January 1975
– "The Lindbergh Kill," Airpower, November 1981

Colonel Charles MacDonald, "Lindbergh in Battle," Collier's, February 1946

John Stanaway, "Satan's Angels," War Monthly, March 1981

Anonymous Documents

SATAN'S ANGELS, Angus and Robertson, 1946

432nd Fighter Squadron Operational Diary

SATAN'S ANGELS 1980 Reunion brochure

Various AAF Southwest Pacific Intelligence Bulletins

Index

Also from the publisher

AMERICAN BOMBER AIRCRAFT VOL.I
Consolidated
B-24 Liberator
John M. & Donna Campbell

Size: 8 1/2" x 11" 256 pages, over 700 b/w and color photos, profiles
ISBN: 0-88740-452-9   hard cover $39.95

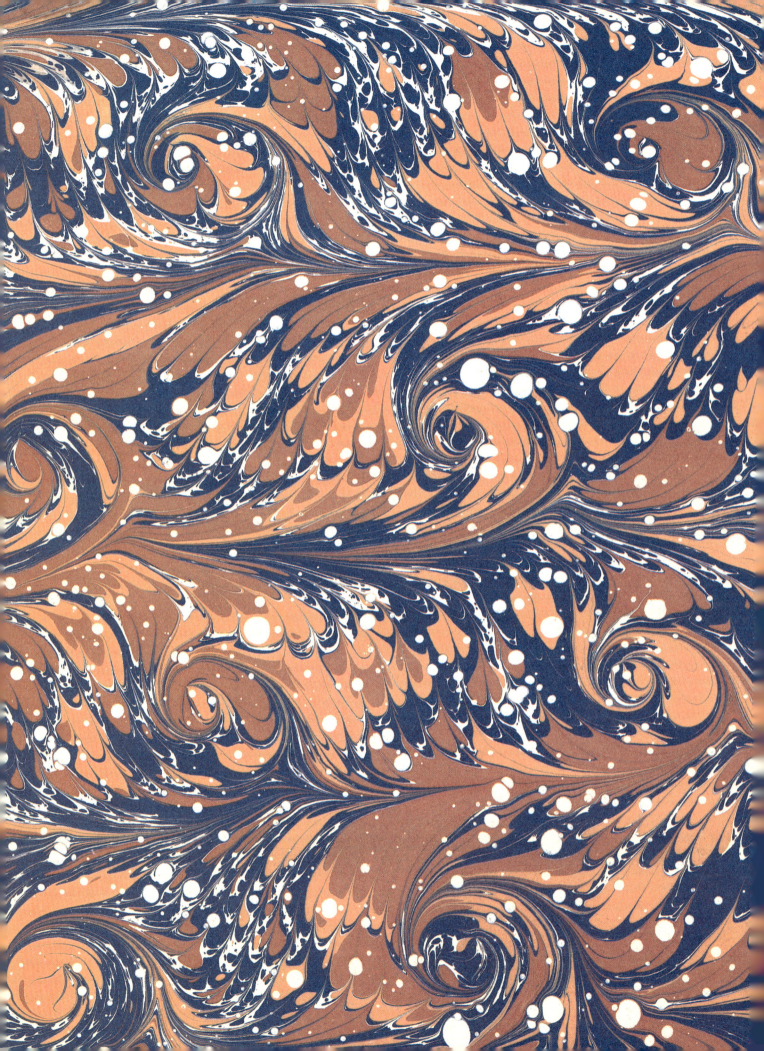